Student Solutions Manual

for use with

Complex Variables and Applications

Seventh Edition

Selected Solutions to Exercises in Chapters 1-7

by

James Ward Brown
Professor of Mathematics
The University of Michigan- Dearborn

Ruel V. Churchill
Late Professor of Mathematics
The University of Michigan

Boston Burr Ridge, IL Dubuque, IA Madison, WI New York San Francisco St. Louis
Bangkok Bogotá Caracas Kuala Lumpur Lisbon London Madrid Mexico City
Milan Montreal New Delhi Santiago Seoul Singapore Sydney Taipei Toronto

The McGraw·Hill Companies

Student Solutions Manual for use with
COMPLEX VARIABLES AND APPLICATIONS, SEVENTH EDITION
JAMES WARD BROWN AND RUEL V. CHURCHILL

Published by McGraw-Hill Higher Education, an imprint of The McGraw-Hill Companies, Inc., 1221 Avenue of the Americas, New York, NY 10020.

1 2 3 4 5 6 7 8 9 0 QPD QPD 0 9 8 7 6 5 4 3

ISBN 0-07-287834-7

www.mhhe.com

Table of Contents

"COMPLEX VARIABLES AND APPLICATIONS" (7/e) by Brown and Churchill

Chapter 1

SECTION 2

1. *(a)* $(\sqrt{2}-i)-i(1-\sqrt{2}i)=\sqrt{2}-i-i-\sqrt{2}=-2i;$

(b) $(2,-3)(-2,1)=(-4+3,6+2)=(-1,8);$

(c) $(3,1)(3,-1)\left(\frac{1}{5},\frac{1}{10}\right)=(10,0)\left(\frac{1}{5},\frac{1}{10}\right)=(2,1).$

2. *(a)* $\operatorname{Re}(iz)=\operatorname{Re}[i(x+iy)]=\operatorname{Re}(-y+ix)=-y=-\operatorname{Im}z;$

(b) $\operatorname{Im}(iz)=\operatorname{Im}[i(x+iy)]=\operatorname{Im}(-y+ix)=x=\operatorname{Re}z.$

3. $(1+z)^2=(1+z)(1+z)=(1+z)\cdot 1+(1+z)z=1\cdot(1+z)+z(1+z)$

$$=1+z+z+z^2=1+2z+z^2.$$

4. If $z=1\pm i$, then $z^2-2z+2=(1\pm i)^2-2(1\pm i)+2=\pm 2i-2\mp 2i+2=0.$

5. To prove that multiplication is commutative, write

$$\begin{aligned} z_1z_2 &= (x_1,y_1)(x_2,y_2)=(x_1x_2-y_1y_2,\ y_1x_2+x_1y_2) \\ &= (x_2x_1-y_2y_1,\ y_2x_1+x_2y_1)=(x_2,y_2)(x_1,y_1)=z_2z_1. \end{aligned}$$

6. *(a)* To verify the associative law for addition, write

$$\begin{aligned} (z_1+z_2)+z_3 &= [(x_1,y_1)+(x_2,y_2)]+(x_3,y_3)=(x_1+x_2,\ y_1+y_2)+(x_3,y_3) \\ &= ((x_1+x_2)+x_3,\ (y_1+y_2)+y_3)=(x_1+(x_2+x_3),\ y_1+(y_2+y_3)) \\ &= (x_1,y_1)+(x_2+x_3,\ y_2+y_3)=(x_1,y_1)+[(x_2,y_2)+(x_3,y_3)] \\ &= z_1+(z_2+z_3). \end{aligned}$$

(b) To verify the distributive law, write

$$\begin{aligned} z(z_1+z_2) &= (x,y)[(x_1,y_1)+(x_2,y_2)] = (x,y)(x_1+x_2,y_1+y_2) \\ &= (xx_1+xx_2-yy_1-yy_2,\ yx_1+yx_2+xy_1+xy_2) \\ &= (xx_1-yy_1+xx_2-yy_2,\ yx_1+xy_1+yx_2+xy_2) \\ &= (xx_1-yy_1,\ yx_1+xy_1)+(xx_2-yy_2,\ yx_2+xy_2) \\ &= (x,y)(x_1,y_1)+(x,y)(x_2,y_2) = zz_1+zz_2. \end{aligned}$$

10. The problem here is to solve the equation $z^2+z+1=0$ for $z=(x,y)$ by writing

$$(x,y)(x,y)+(x,y)+(1,0)=(0,0).$$

Since

$$(x^2-y^2+x+1,\ 2xy+y)=(0,0),$$

it follows that

$$x^2-y^2+x+1=0 \quad \text{and} \quad 2xy+y=0.$$

By writing the second of these equations as $(2x+1)y=0$, we see that either $2x+1=0$ or $y=0$. If $y=0$, the first equation becomes $x^2+x+1=0$, which has no real roots (according to the quadratic formula). Hence $2x+1=0$, or $x=-1/2$. In that case, the first equation reveals that $y^2=3/4$, or $y=\pm\sqrt{3}/2$. Thus

$$z=(x,y)=\left(-\frac{1}{2},\pm\frac{\sqrt{3}}{2}\right).$$

SECTION 3

1. *(a)* $\dfrac{1+2i}{3-4i}+\dfrac{2-i}{5i}=\dfrac{(1+2i)(3+4i)}{(3-4i)(3+4i)}+\dfrac{(2-i)(-5i)}{(5i)(-5i)}=\dfrac{-5+10i}{25}+\dfrac{-5-10i}{25}=-\dfrac{2}{5};$

(b) $\dfrac{5i}{(1-i)(2-i)(3-i)}=\dfrac{5i}{(1-3i)(3-i)}=\dfrac{5i}{-10i}=-\dfrac{1}{2};$

(c) $(1-i)^4=[(1-i)(1-i)]^2=(-2i)^2=-4.$

2. *(a)* $(-1)z=-z$ since $z+(-1)z=z[1+(-1)]=z\cdot 0=0;$

(b) $\dfrac{1}{1/z}=\dfrac{1}{z^{-1}}\cdot\dfrac{z}{z}=\dfrac{z}{1}=z \quad (z\neq 0).$

3. $(z_1z_2)(z_3z_4) = z_1[z_2(z_3z_4)] = z_1[(z_2z_3)z_4] = z_1[(z_3z_2)z_4)] = z_1[z_3(z_2z_4)] = (z_1z_3)(z_2z_4).$

6. $$\frac{z_1z_2}{z_3z_4} = z_1z_2\left(\frac{1}{z_3z_4}\right) = z_1z_2\left(\frac{1}{z_3}\right)\left(\frac{1}{z_4}\right) = z_1\left(\frac{1}{z_3}\right)z_2\left(\frac{1}{z_4}\right) = \left(\frac{z_1}{z_3}\right)\left(\frac{z_2}{z_4}\right) \qquad (z_3 \neq 0, z_4 \neq 0).$$

7. $$\frac{z_1z}{z_2z} = \left(\frac{z_1}{z_2}\right)\left(\frac{z}{z}\right) = \left(\frac{z_1}{z_2}\right)z\left(\frac{1}{z}\right) = \left(\frac{z_1}{z_2}\right)(zz^{-1}) = \left(\frac{z_1}{z_2}\right)\cdot 1 = \frac{z_1}{z_2} \qquad (z_2 \neq 0, z \neq 0).$$

SECTION 4

1. *(a)* $z_1 = 2i, \quad z_2 = \dfrac{2}{3} - i$

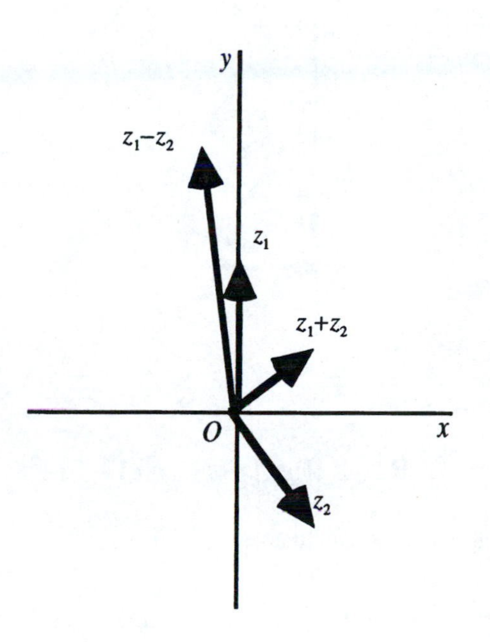

(b) $z_1 = (-\sqrt{3}, 1), \quad z_2 = (\sqrt{3}, 0)$

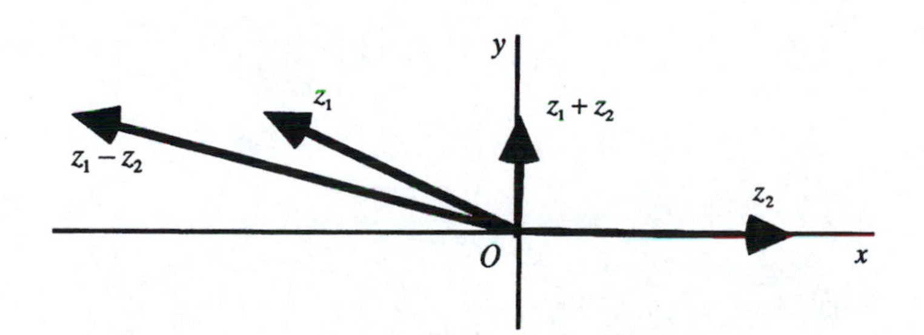

(c) $z_1 = (-3,1), \quad z_2 = (1,4)$

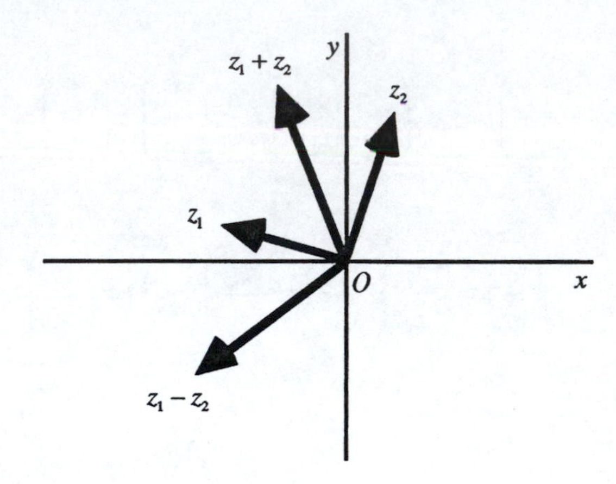

(d) $z_1 = x_1 + iy_1, \quad z_2 = x_1 - iy_1$

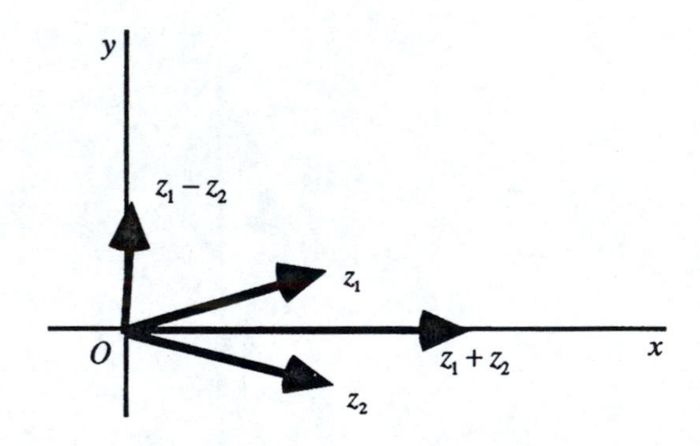

2. Inequalities (3), Sec. 4, are

$$\operatorname{Re} z \le |\operatorname{Re} z| \le |z| \quad \text{and} \quad \operatorname{Im} z \le |\operatorname{Im} z| \le |z|.$$

These are obvious if we write them as

$$x \le |x| \le \sqrt{x^2 + y^2} \quad \text{and} \quad y \le |y| \le \sqrt{x^2 + y^2}.$$

3. In order to verify the inequality $\sqrt{2}|z| \ge |\operatorname{Re} z| + |\operatorname{Im} z|$, we rewrite it in the following ways:

$$\sqrt{2}\sqrt{x^2 + y^2} \ge |x| + |y|,$$

$$2(x^2 + y^2) \ge |x|^2 + 2|x||y| + |y|^2,$$

$$|x|^2 - 2|x||y| + |y|^2 \ge 0,$$

$$(|x| - |y|)^2 \ge 0.$$

This last form of the inequality to be verified is obviously true since the left-hand side is a perfect square.

4. *(a)* Rewrite $|z-1+i|=1$ as $|z-(1-i)|=1$. This is the circle centered at $1-i$ with radius 1. It is shown below.

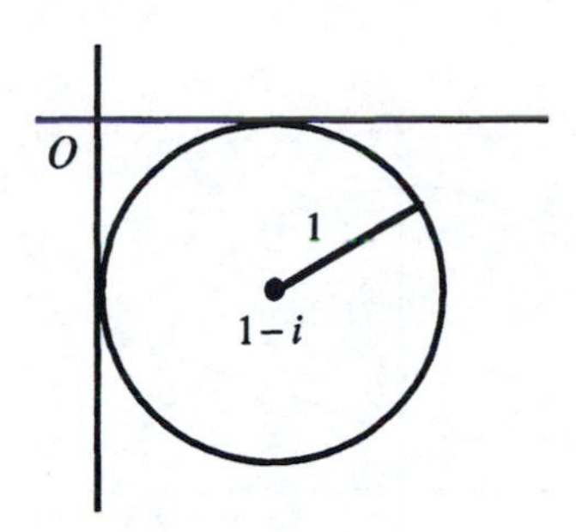

5. *(a)* Write $|z-4i|+|z+4i|=10$ as $|z-4i|+|z-(-4i)|=10$ to see that this is the locus of all points z such that the sum of the distances from z to $4i$ and $-4i$ is a constant. Such a curve is an ellipse with foci $\pm 4i$.

(b) Write $|z-1|=|z+i|$ as $|z-1|=|z-(-i)|$ to see that this is the locus of all points z such that the distance from z to 1 is always the same as the distance to $-i$. The curve is, then, the perpendicular bisector of the line segment from 1 to $-i$.

SECTION 5

1. *(a)* $\overline{z+3i}=\overline{\overline{z}}+\overline{3i}=z-3i;$

(b) $\overline{iz}=\overline{i}\,\overline{z}=-i\overline{z};$

(c) $\overline{(2+i)^2}=\left(\overline{2+i}\right)^2=(2-i)^2=4-4i+i^2=4-4i-1=3-4i;$

(d) $|(2\overline{z}+5)(\sqrt{2}-i)|=|2\overline{z}+5||\sqrt{2}-i|=|\overline{2z+5}|\sqrt{2+1}=\sqrt{3}\,|2z+5|.$

2. *(a)* Rewrite $\text{Re}(\overline{z}-i)=2$ as $\text{Re}[x+i(-y-1)]=2$, or $x=2$. This is the vertical line through the point $z=2$, shown below.

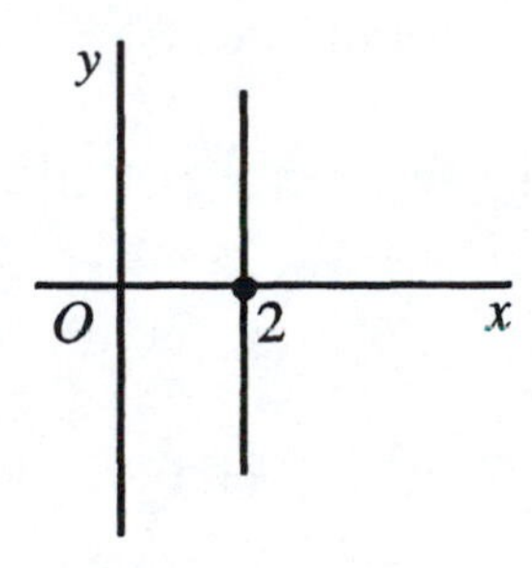

(b) Rewrite $|2z-i|=4$ as $2\left|z-\frac{i}{2}\right|=4$, or $\left|z-\frac{i}{2}\right|=2$. This is the circle centered at $\frac{i}{2}$ with radius 2, shown below.

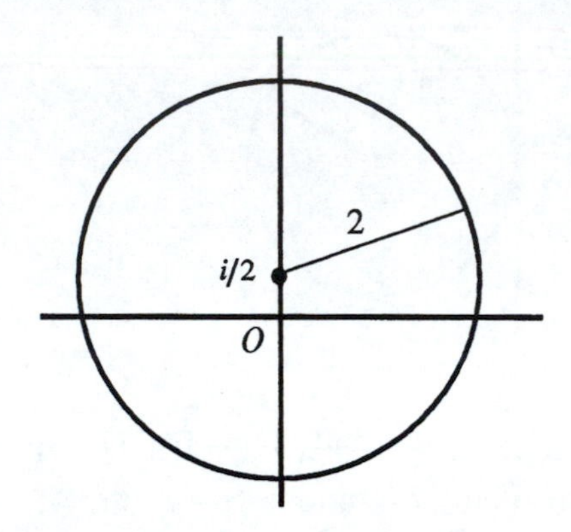

3. Write $z_1 = x_1 + iy_1$ and $z_2 = x_2 + iy_2$. Then

$$\overline{z_1 - z_2} = \overline{(x_1 + iy_1) - (x_2 + iy_2)} = \overline{(x_1 - x_2) + i(y_1 - y_2)}$$
$$= (x_1 - x_2) - i(y_1 - y_2) = (x_1 - iy_1) - (x_2 - iy_2) = \bar{z}_1 - \bar{z}_2$$

and

$$\overline{z_1 z_2} = \overline{(x_1 + iy_1)(x_2 + iy_2)} = \overline{(x_1x_2 - y_1y_2) + i(y_1x_2 + x_1y_2)}$$
$$= (x_1x_2 - y_1y_2) - i(y_1x_2 + x_1y_2) = (x_1 - iy_1)(x_2 - iy_2) = \bar{z}_1\bar{z}_2.$$

4. *(a)* $\overline{z_1z_2z_3} = \overline{(z_1z_2)z_3} = \overline{z_1z_2}\,\bar{z}_3 = (\bar{z}_1\,\bar{z}_2)\bar{z}_3 = \bar{z}_1\,\bar{z}_2\,\bar{z}_3$;

(b) $\overline{z^4} = \overline{z^2z^2} = \overline{z^2}\,\overline{z^2} = \overline{zz}\,\overline{zz} = (\bar{z}\,\bar{z})(\bar{z}\,\bar{z}) = \bar{z}\bar{z}\bar{z}\bar{z} = \bar{z}^{\,4}$.

6. *(a)* $\overline{\left(\dfrac{z_1}{z_2z_3}\right)} = \dfrac{\bar{z}_1}{\overline{z_2z_3}} = \dfrac{\bar{z}_1}{\bar{z}_2\,\bar{z}_3}$;

(b) $\left|\dfrac{z_1}{z_2z_3}\right| = \dfrac{|z_1|}{|z_2z_3|} = \dfrac{|z_1|}{|z_2||z_3|}$.

8. In this problem, we shall use the inequalities (see Sec. 4)

$$|\operatorname{Re} z| \le |z| \quad \text{and} \quad |z_1 + z_2 + z_3| \le |z_1| + |z_2| + |z_3|.$$

Specifically, when $|z| \le 1$,

$$|\operatorname{Re}(2 + \bar{z} + z^3)| \le |2 + \bar{z} + z^3| \le 2 + |\bar{z}| + |z^3| = 2 + |z| + |z|^3 \le 2 + 1 + 1 = 4.$$

10. First write $z^4-4z^2+3=(z^2-1)(z^2-3)$. Then observe that when $|z|=2$,

$$|z^2-1|\ge\left||z^2|-|1|\right|=\left||z|^2-1\right|=|4-1|=3$$

and

$$|z^2-3|\ge\left||z^2|-|3|\right|=\left||z|^2-3\right|=|4-3|=1.$$

Thus, when $|z|=2$,

$$|z^4-4z^2+3|=|z^2-1|\cdot|z^2-3|\ge 3\cdot 1=3.$$

Consequently, when z lies on the circle $|z|=2$,

$$\left|\frac{1}{z^4-4z^2+3}\right|=\frac{1}{|z^4-4z^2+3|}\le\frac{1}{3}.$$

11. *(a)* Prove that z is real $\Leftrightarrow$ $\bar{z}=z$.

($\Leftarrow$) Suppose that $\bar{z}=z$, so that $x-iy=x+iy$. This means that $i2y=0$, or $y=0$. Thus $z=x+i0=x$, or z is real.

($\Rightarrow$) Suppose that z is real, so that $z=x+i0$. Then $\bar{z}=x-i0=x+i0=z$.

(b) Prove that z is either real or pure imaginary $\Leftrightarrow$ $\bar{z}^2=z^2$.

($\Leftarrow$) Suppose that $\bar{z}^2=z^2$. Then $(x-iy)^2=(x+iy)^2$, or $i4xy=0$. But this can be only if either $x=0$ or $y=0$, or possibly $x=y=0$. Thus z is either real or pure imaginary.

($\Rightarrow$) Suppose that z is either real or pure imaginary. If z is real, so that $z=x$, then $\bar{z}^2=x^2=z^2$. If z is pure imaginary, so that $z=iy$, then $\bar{z}^2=(-iy)^2=(iy)^2=z^2$.

12. *(a)* We shall use mathematical induction to show that

$$\overline{z_1+z_2+\cdots+z_n}=\bar{z}_1+\bar{z}_2+\cdots+\bar{z}_n \qquad (n=2,3,\ldots).$$

This is known when $n=2$ (Sec. 5). Assuming now that it is true when $n=m$, we may write

$$\begin{aligned}\overline{z_1+z_2+\cdots+z_m+z_{m+1}}&=\overline{(z_1+z_2+\cdots+z_m)+z_{m+1}}\\&=\overline{(z_1+z_2+\cdots+z_m)}+\bar{z}_{m+1}\\&=\bar{z}_1+\bar{z}_2+\cdots+\bar{z}_m)+\bar{z}_{m+1}\\&=\bar{z}_1+\bar{z}_2+\cdots+\bar{z}_m+\bar{z}_{m+1}.\end{aligned}$$

(b) In the same way, we can show that

$$\overline{z_1 z_2 \cdots z_n} = \overline{z}_1\,\overline{z}_2 \cdots \overline{z}_n \qquad (n = 2, 3, \ldots).$$

This is true when $n = 2$ (Sec. 5). Assuming that it is true when $n = m$, we write

$$\overline{z_1 z_2 \cdots z_m z_{m+1}} = \overline{(z_1 z_2 \cdots z_m) z_{m+1}} = \overline{(z_1 z_2 \cdots z_m)}\;\overline{z}_{m+1}$$
$$= (\overline{z}_1 \overline{z}_2 \cdots \overline{z}_m)\overline{z}_{m+1} = \overline{z}_1 \overline{z}_2 \cdots \overline{z}_m \overline{z}_{m+1}.$$

14. The identities (Sec. 5) $z\overline{z} = |z|^2$ and $\operatorname{Re} z = \dfrac{z + \overline{z}}{2}$ enable us to write $|z - z_0| = R$ as

$$(z - z_0)(\overline{z} - \overline{z}_0) = R^2,$$

$$z\overline{z} - (z\overline{z}_0 + \overline{z\overline{z}_0}) + z_0\overline{z}_0 = R^2,$$

$$|z|^2 - 2\operatorname{Re}(z\overline{z}_0) + |z_0|^2 = R^2.$$

15. Since $x = \dfrac{z + \overline{z}}{2}$ and $y = \dfrac{z - \overline{z}}{2i}$, the hyperbola $x^2 - y^2 = 1$ can be written in the following ways:

$$\left(\frac{z + \overline{z}}{2}\right)^2 - \left(\frac{z - \overline{z}}{2i}\right)^2 = 1,$$

$$\frac{z^2 + 2z\overline{z} + \overline{z}^2}{4} + \frac{z^2 - 2z\overline{z} + \overline{z}^2}{4} = 1,$$

$$\frac{2z^2 + 2\overline{z}^2}{4} = 1,$$

$$z^2 + \overline{z}^2 = 2.$$

SECTION 7

1. *(a)* Since

$$\arg\left(\frac{i}{-2 - 2i}\right) = \arg i - \arg(-2 - 2i),$$

one value of $\arg\left(\dfrac{i}{-2 - 2i}\right)$ is $\dfrac{\pi}{2} - \left(-\dfrac{3\pi}{4}\right)$, or $\dfrac{5\pi}{4}$. Consequently, the principal value is

$$\frac{5\pi}{4} - 2\pi, \text{ or } -\frac{3\pi}{4}.$$

(b) Since

$$\arg(\sqrt{3}-i)^6 = 6\arg(\sqrt{3}-i),$$

one value of $\arg(\sqrt{3}-i)^6$ is $6\left(-\dfrac{\pi}{6}\right)$, or $-\pi$. So the principal value is $-\pi+2\pi$, or π.

4. The solution $\theta=\pi$ of the equation $|e^{i\theta}-1|=2$ in the interval $0\le\theta<2\pi$ is geometrically evident if we recall that $e^{i\theta}$ lies on the circle $|z|=1$ and that $|e^{i\theta}-1|$ is the distance between the points $e^{i\theta}$ and 1. See the figure below.

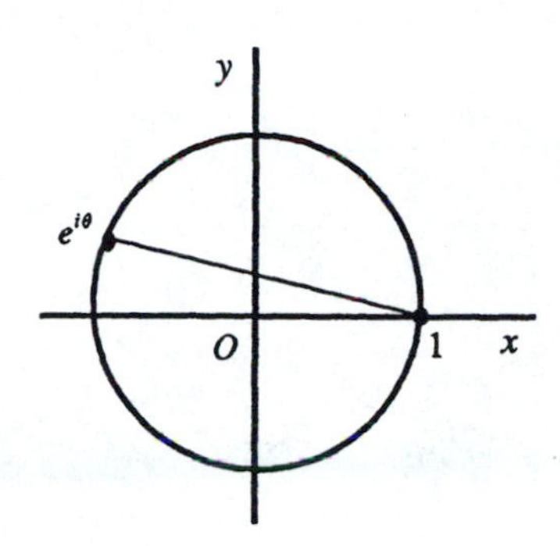

5. We know from de Moivre's formula that

$$(\cos\theta + i\sin\theta)^3 = \cos 3\theta + i\sin 3\theta,$$

or

$$\cos^3\theta + 3\cos^2\theta(i\sin\theta) + 3\cos\theta(i\sin\theta)^2 + (i\sin\theta)^3 = \cos 3\theta + i\sin 3\theta.$$

That is,

$$(\cos^3\theta - 3\cos\theta\sin^2\theta) + i(3\cos^2\theta\sin\theta - \sin^3\theta) = \cos 3\theta + i\sin 3\theta.$$

By equating real parts and then imaginary parts here, we arrive at the desired trigonometric identities:

(a) $\cos 3\theta = \cos^3\theta - 3\cos\theta\sin^2\theta$; *(b)* $\sin 3\theta = 3\cos^2\theta\sin\theta - \sin^3\theta$.

8. Here $z=re^{i\theta}$ is any nonzero complex number and n a negative integer $(n=-1,-2,\ldots)$. Also, $m=-n=1,2,\ldots$. By writing

$$(z^m)^{-1} = (r^m e^{im\theta})^{-1} = \frac{1}{r^m}e^{i(-m\theta)}$$

and

$$(z^{-1})^m = \left[\frac{1}{r}e^{i(-\theta)}\right]^m = \left(\frac{1}{r}\right)^m e^{i(-m\theta)} = \frac{1}{r^m}e^{i(-m\theta)},$$

we see that $(z^m)^{-1}=(z^{-1})^m$. Thus the definition $z^n=(z^{-1})^m$ can also be written as $z^n=(z^m)^{-1}$.

9. First of all, given two nonzero complex numbers z_1 and z_2, suppose that there are complex numbers c_1 and c_2 such that $z_1 = c_1c_2$ and $z_2 = c_1\overline{c}_2$. Since

$$|z_1| = |c_1||c_2| \quad \text{and} \quad |z_2| = |c_1||\overline{c_2}| = |c_1||c_2|,$$

it follows that $|z_1| = |z_2|$.

Suppose, on the other hand, that we know only that $|z_1| = |z_2|$. We may write

$$z_1 = r_1 \exp(i\theta_1) \quad \text{and} \quad z_2 = r_1 \exp(i\theta_2).$$

If we introduce the numbers

$$c_1 = r_1 \exp\left(i\frac{\theta_1 + \theta_2}{2}\right) \quad \text{and} \quad c_2 = \exp\left(i\frac{\theta_1 - \theta_2}{2}\right),$$

we find that

$$c_1c_2 = r_1 \exp\left(i\frac{\theta_1 + \theta_2}{2}\right)\exp\left(i\frac{\theta_1 - \theta_2}{2}\right) = r_1 \exp(i\theta_1) = z_1$$

and

$$c_1\overline{c}_2 = r_1 \exp\left(i\frac{\theta_1 + \theta_2}{2}\right)\exp\left(-i\frac{\theta_1 - \theta_2}{2}\right) = r_1 \exp\theta_2 = z_2.$$

That is,

$$z_1 = c_1c_2 \quad \text{and} \quad z_2 = c_1\overline{c}_2.$$

10. If $S = 1 + z + z^2 + \cdots + z^n$, then

$$S - zS = (1 + z + z^2 + \cdots + z^n) - (z + z^2 + z^3 + \cdots + z^{n+1}) = 1 - z^{n+1}.$$

Hence $S = \dfrac{1 - z^{n+1}}{1 - z}$, provided $z \neq 1$. That is,

$$1 + z + z^2 + \cdots + z^n = \frac{1 - z^{n+1}}{1 - z} \qquad (z \neq 1).$$

Putting $z = e^{i\theta}$ $(0 < \theta < 2\pi)$ in this identity, we have

$$1 + e^{i\theta} + e^{i2\theta} + \cdots + e^{in\theta} = \frac{1 - e^{i(n+1)\theta}}{1 - e^{i\theta}}.$$

Now the real part of the left-hand side here is evidently

$$1+\cos\theta+\cos 2\theta+\cdots+\cos n\theta;$$

and, to find the real part of the right-hand side, we write that side in the form

$$\frac{1-\exp[i(n+1)\theta]}{1-\exp(i\theta)}\cdot\frac{\exp\left(-i\frac{\theta}{2}\right)}{\exp\left(-i\frac{\theta}{2}\right)}=\frac{\exp\left(-i\frac{\theta}{2}\right)-\exp\left[i\frac{(2n+1)\theta}{2}\right]}{\exp\left(-i\frac{\theta}{2}\right)-\exp\left(i\frac{\theta}{2}\right)},$$

which becomes

$$\frac{\cos\frac{\theta}{2}-i\sin\frac{\theta}{2}-\cos\frac{(2n+1)\theta}{2}-i\sin\frac{(2n+1)\theta}{2}}{-2i\sin\frac{\theta}{2}}\cdot\frac{i}{i},$$

or

$$\frac{\left[\sin\frac{\theta}{2}+\sin\frac{(2n+1)\theta}{2}\right]+i\left[\cos\frac{\theta}{2}-\cos\frac{(2n+1)\theta}{2}\right]}{2\sin\frac{\theta}{2}}.$$

The real part of this is clearly

$$\frac{1}{2}+\frac{\sin\frac{(2n+1)\theta}{2}}{2\sin\frac{\theta}{2}},$$

and we arrive at *Lagrange's trigonometric identity*:

$$1+\cos\theta+\cos 2\theta+\cdots+\cos n\theta=\frac{1}{2}+\frac{\sin\frac{(2n+1)\theta}{2}}{2\sin\frac{\theta}{2}}\qquad(0<\theta<2\pi).$$

SECTION 9

1. *(a)* Since $2i = 2\exp\left[i\left(\frac{\pi}{2} + 2k\pi\right)\right]$ $(k = 0, \pm 1, \pm 2, \ldots)$, the desired roots are

$$(2i)^{1/2} = \sqrt{2}\exp\left[i\left(\frac{\pi}{4} + k\pi\right)\right] \qquad (k = 0, 1).$$

That is,

$$c_0 = \sqrt{2}e^{i\pi/4} = \sqrt{2}\left(\cos\frac{\pi}{4} + i\sin\frac{\pi}{4}\right) = \sqrt{2}\left(\frac{1}{\sqrt{2}} + \frac{i}{\sqrt{2}}\right) = 1 + i$$

and

$$c_1 = (\sqrt{2}e^{i\pi/4})e^{i\pi} = -c_0 = -(1 + i),$$

c_0 being the principal root. These are sketched below.

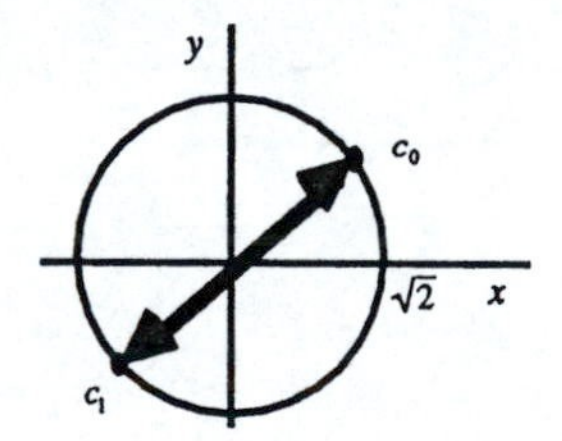

(b) Observe that $1 - \sqrt{3}i = 2\exp\left[i\left(-\frac{\pi}{3} + 2k\pi\right)\right]$ $(k = 0, \pm 1, \pm 2, \ldots)$. Hence

$$(1 - \sqrt{3}i)^{1/2} = \sqrt{2}\exp\left[i\left(-\frac{\pi}{6} + k\pi\right)\right] \qquad (k = 0, 1).$$

The principal root is

$$c_0 = \sqrt{2}e^{-i\pi/6} = \sqrt{2}\left(\cos\frac{\pi}{6} - i\sin\frac{\pi}{6}\right) = \sqrt{2}\left(\frac{\sqrt{3}}{2} - \frac{i}{2}\right) = \frac{\sqrt{3} - i}{\sqrt{2}},$$

and the other root is

$$c_1 = (\sqrt{2}e^{-i\pi/6})e^{i\pi} = -c_0 = -\frac{\sqrt{3} - i}{\sqrt{2}}.$$

These roots are shown below.

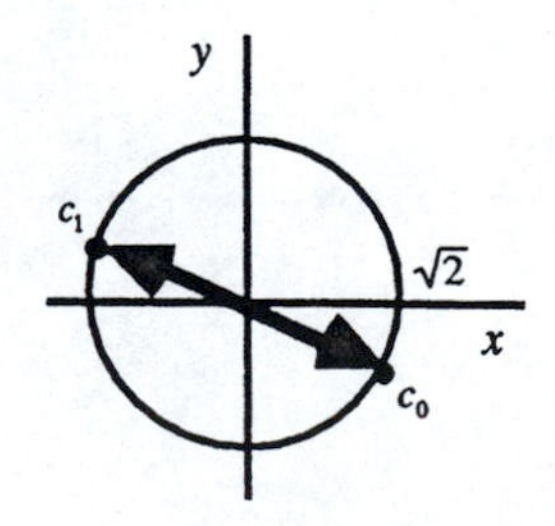

2. *(a)* Since $-16 = 16\exp[i(\pi + 2k\pi)]$ $(k = 0, \pm 1, \pm 2, \ldots)$, the needed roots are

$$(-16)^{1/4} = 2\exp\left[i\left(\frac{\pi}{4} + \frac{k\pi}{2}\right)\right] \qquad (k = 0,1,2,3).$$

The principal root is

$$c_0 = 2e^{i\pi/4} = 2\left(\cos\frac{\pi}{4} + i\sin\frac{\pi}{4}\right) = 2\left(\frac{1}{\sqrt{2}} + \frac{i}{\sqrt{2}}\right) = \sqrt{2}(1+i).$$

The other three roots are

$$c_1 = (2e^{i\pi/4})e^{i\pi/2} = c_0 i = \sqrt{2}(1+i)i = -\sqrt{2}(1-i),$$

$$c_2 = (2e^{i\pi/4})e^{i\pi} = -c_0 = -\sqrt{2}(1+i),$$

and

$$c_3 = (2e^{i\pi/4})e^{i3\pi/2} = c_0(-i) = \sqrt{2}(1+i)(-i) = \sqrt{2}(1-i).$$

The four roots are shown below.

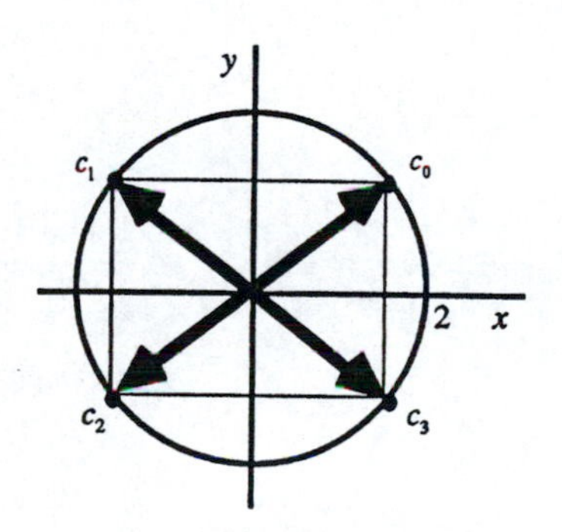

(b) First write $-8 - 8\sqrt{3}i = 16\exp\left[i\left(-\frac{2\pi}{3} + 2k\pi\right)\right]$ $(k = 0, \pm 1, \pm 2, \ldots)$. Then

$$(-8 - 8\sqrt{3}i)^{1/4} = 2\exp\left[i\left(-\frac{\pi}{6} + \frac{k\pi}{2}\right)\right] \qquad (k = 0,1,2,3).$$

The principal root is

$$c_0 = 2e^{-i\pi/6} = 2\left(\cos\frac{\pi}{6} - i\sin\frac{\pi}{6}\right) = 2\left(\frac{\sqrt{3}}{2} - \frac{i}{2}\right) = \sqrt{3} - i.$$

The others are

$$c_1 = (2e^{-i\pi/6})e^{i\pi/2} = c_0 i = 1 + \sqrt{3}i,$$

$$c_2 = (2e^{-i\pi/6})e^{i\pi} = -c_0 = -(\sqrt{3} - i),$$

$$c_3 = (2e^{-i\pi/6})e^{i3\pi/2} = c_0(-i) = -(1 + \sqrt{3}i).$$

These roots are all shown below.

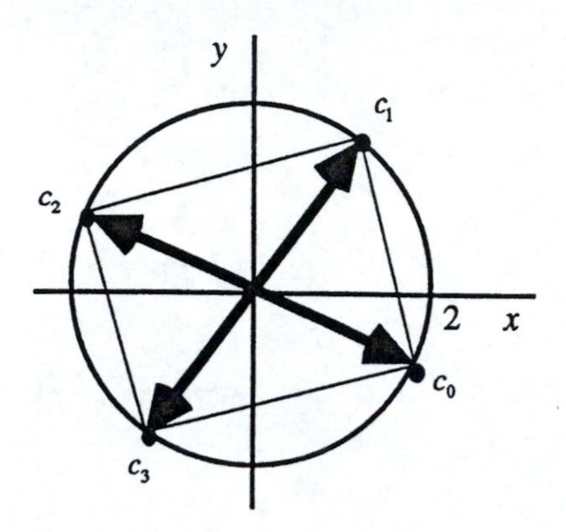

3. *(a)* By writing $-1 = 1\exp[i(\pi + 2k\pi)]$ $(k = 0, \pm1, \pm2, \ldots)$, we see that

$$(-1)^{1/3} = \exp\left[i\left(\frac{\pi}{3} + \frac{2k\pi}{3}\right)\right] \qquad (k = 0,1,2).$$

The principal root is

$$c_0 = e^{i\pi/3} = \cos\frac{\pi}{3} + i\sin\frac{\pi}{3} = \frac{1 + \sqrt{3}i}{2}.$$

The other two roots are

$$c_1 = e^{i\pi} = -1$$

and

$$c_2 = e^{i5\pi/3} = e^{i2\pi}e^{-i\pi/3} = \cos\frac{\pi}{3} - i\sin\frac{\pi}{3} = \frac{1 - \sqrt{3}i}{2}.$$

All three roots are shown below.

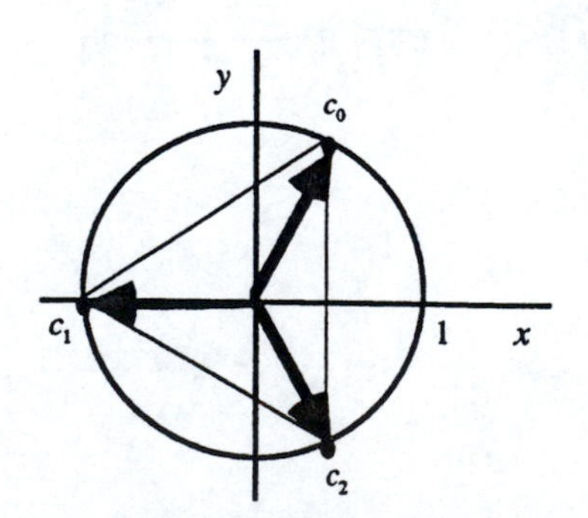

(b) Since $8 = 8\exp[i(0+2k\pi)]$ $(k = 0, \pm 1, \pm 2, \ldots)$, the desired roots of 8 are

$$8^{1/6} = \sqrt{2}\exp\left(i\frac{k\pi}{3}\right) \qquad (k = 0,1,2,3,4,5),$$

the principal one being

$$c_0 = \sqrt{2}.$$

The others are

$$c_1 = \sqrt{2}e^{i\pi/3} = \sqrt{2}\left(\cos\frac{\pi}{3} + i\sin\frac{\pi}{3}\right) = \sqrt{2}\left(\frac{1}{2} + \frac{\sqrt{3}}{2}i\right) = \frac{1+\sqrt{3}i}{\sqrt{2}},$$

$$c_2 = (\sqrt{2}e^{-i\pi/3})e^{i\pi} = \sqrt{2}\left(\cos\frac{\pi}{3} - i\sin\frac{\pi}{3}\right)(-1) = -\sqrt{2}\left(\frac{1}{2} - \frac{\sqrt{3}}{2}i\right) = -\frac{1-\sqrt{3}i}{\sqrt{2}},$$

$$c_3 = \sqrt{2}e^{i\pi} = -\sqrt{2},$$

$$c_4 = (\sqrt{2}e^{i\pi/3})e^{i\pi} = -c_1 = -\frac{1+\sqrt{3}i}{\sqrt{2}},$$

and

$$c_5 = (\sqrt{2}e^{i2\pi/3})e^{i\pi} = -c_2 = \frac{1-\sqrt{3}i}{\sqrt{2}}.$$

All six roots are shown below.

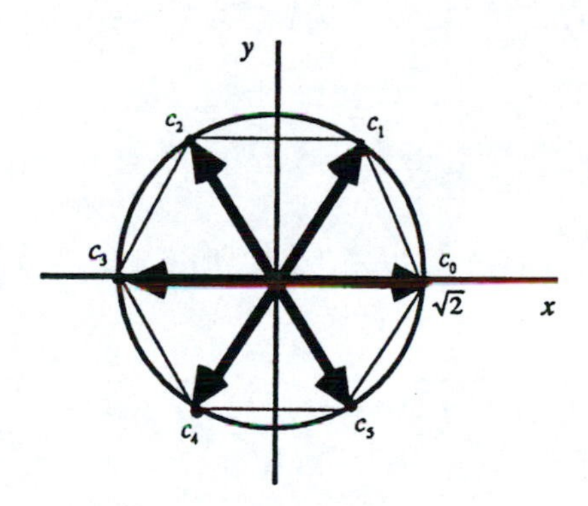

4. The three cube roots of the number $z_0 = -4\sqrt{2} + 4\sqrt{2}i = 8\exp\left(i\frac{3\pi}{4}\right)$ are evidently

$$(z_0)^{1/3} = 2\exp\left[i\left(\frac{\pi}{4} + \frac{2k\pi}{3}\right)\right] \qquad (k = 0,1,2).$$

In particular,

$$c_0 = 2\exp\left(i\frac{\pi}{4}\right) = \sqrt{2}(1+i).$$

With the aid of the number $\omega_3 = \dfrac{-1+\sqrt{3}i}{2}$, we obtain the other two roots:

$$c_1 = c_0\omega_3 = \sqrt{2}(1+i)\left(\frac{-1+\sqrt{3}i}{2}\right) = \frac{-(\sqrt{3}+1)+(\sqrt{3}-1)i}{\sqrt{2}},$$

$$c_2 = c_0\omega_3^2 = (c_0\omega_3)\omega_3 = \left[\frac{-(\sqrt{3}+1)+(\sqrt{3}-1)i}{\sqrt{2}}\right]\left(\frac{-1+\sqrt{3}i}{2}\right) = \frac{(\sqrt{3}-1)-(\sqrt{3}+1)i}{\sqrt{2}}.$$

5. *(a)* Let a denote any fixed real number. In order to find the two square roots of $a+i$ in exponential form, we write

$$A = |a+i| = \sqrt{a^2+1} \quad \text{and} \quad \alpha = \operatorname{Arg}(a+i).$$

Since

$$a+i = A\exp[i(\alpha+2k\pi)] \qquad (k = 0,\pm1,\pm2,\ldots),$$

we see that

$$(a+i)^{1/2} = \sqrt{A}\exp\left[i\left(\frac{\alpha}{2}+k\pi\right)\right] \qquad (k=0,1).$$

That is, the desired square roots are

$$\sqrt{A}e^{i\alpha/2} \quad \text{and} \quad \sqrt{A}e^{i\alpha/2}e^{i\pi} = -\sqrt{A}e^{i\alpha/2}.$$

(b) Since $a+i$ lies above the real axis, we know that $0 < \alpha < \pi$. Thus $0 < \dfrac{\alpha}{2} < \dfrac{\pi}{2}$, and this tells us that $\cos\left(\dfrac{\alpha}{2}\right) > 0$ and $\sin\left(\dfrac{\alpha}{2}\right) > 0$. Since $\cos\alpha = \dfrac{a}{A}$, it follows that

$$\cos\frac{\alpha}{2} = \sqrt{\frac{1+\cos\alpha}{2}} = \frac{1}{\sqrt{2}}\sqrt{1+\frac{a}{A}} = \frac{\sqrt{A+a}}{\sqrt{2}\sqrt{A}}$$

and

$$\sin\frac{\alpha}{2} = \sqrt{\frac{1-\cos\alpha}{2}} = \frac{1}{\sqrt{2}}\sqrt{1-\frac{a}{A}} = \frac{\sqrt{A-a}}{\sqrt{2}\sqrt{A}}.$$

Consequently,

$$\pm\sqrt{A}e^{i\alpha/2} = \pm\sqrt{A}\left(\cos\frac{\alpha}{2}+i\sin\frac{\alpha}{2}\right) = \pm\sqrt{A}\left(\frac{\sqrt{A+a}}{\sqrt{2}\sqrt{A}}+i\frac{\sqrt{A-a}}{\sqrt{2}\sqrt{A}}\right)$$
$$= \pm\frac{1}{\sqrt{2}}(\sqrt{A+a}+i\sqrt{A-a}).$$

6. The four roots of the equation $z^4 + 4 = 0$ are the four fourth roots of the number -4. To find those roots, we write $-4 = 4\exp[i(\pi + 2k\pi)]$ $(k = 0, \pm 1, \pm 2, \ldots)$. Then

$$(-4)^{1/4} = \sqrt{2}\exp\left[i\left(\frac{\pi}{4} + \frac{k\pi}{2}\right)\right] = \sqrt{2}e^{i\pi/4}e^{ik\pi/2} \qquad (k = 0,1,2,3).$$

To be specific,

$$c_0 = \sqrt{2}e^{i\pi/4} = \sqrt{2}\left(\cos\frac{\pi}{4} + i\sin\frac{\pi}{4}\right) = \sqrt{2}\left(\frac{1}{\sqrt{2}} + i\frac{1}{\sqrt{2}}\right) = 1 + i,$$

$$c_1 = c_0 e^{i\pi/2} = (1+i)i = -1 + i,$$

$$c_2 = c_0 e^{i\pi} = (1+i)(-1) = -1 - i,$$

$$c_3 = c_0 e^{i3\pi/2} = (1+i)(-i) = 1 - i.$$

This enables us to write

$$\begin{aligned} z^4 + 4 &= (z - c_0)(z - c_1)(z - c_2)(z - c_3) \\ &= [(z - c_1)(z - c_2)] \cdot [(z - c_0)(z - c_3)] \\ &= [(z+1) - i][(z+1) + i] \cdot [(z-1) - i][(z-1) + i] \\ &= \left[(z+1)^2 + 1\right] \cdot \left[(z-1)^2 + 1\right] \\ &= \left(z^2 + 2z + 2\right)\left(z^2 - 2z + 2\right). \end{aligned}$$

7. Let c be any nth root of unity other than unity itself. With the aid of the identity (see Exercise 10, Sec. 7),

$$1 + z + z^2 + \cdots + z^{n-1} = \frac{1 - z^n}{1 - z} \qquad (z \neq 1),$$

we find that

$$1 + c + c^2 + \cdots + c^{n-1} = \frac{1 - c^n}{1 - c} = \frac{1 - 1}{1 - c} = 0.$$

9. Observe first that

$$(z^{1/m})^{-1} = \left[\sqrt[m]{r}\exp\frac{i(\theta + 2k\pi)}{m}\right]^{-1} = \frac{1}{\sqrt[m]{r}}\exp\frac{i(-\theta - 2k\pi)}{m} = \frac{1}{\sqrt[m]{r}}\exp\frac{i(-\theta)}{m}\exp\frac{i(-2k\pi)}{m}$$

and

$$(z^{-1})^{1/m} = \sqrt[m]{\frac{1}{r}}\exp\frac{i(-\theta+2k\pi)}{m} = \frac{1}{\sqrt[m]{r}}\exp\frac{i(-\theta)}{m}\exp\frac{i(2k\pi)}{m},$$

where $k = 0,1,2,\ldots,m-1$. Since the set

$$\exp\frac{i(-2k\pi)}{m} \qquad (k = 0,1,2,\ldots,m-1)$$

is the same as the set

$$\exp\frac{i(2k\pi)}{m} \qquad (k = 0,1,2,\ldots,m-1),$$

but in reverse order, we find that $(z^{1/m})^{-1} = (z^{-1})^{1/m}$.

SECTION 10

1. *(a)* Write $|z-2+i| \le 1$ as $|z-(2-i)| \le 1$ to see that this is the set of points inside and on the circle centered at the point $2-i$ with radius 1. It is *not* a domain.

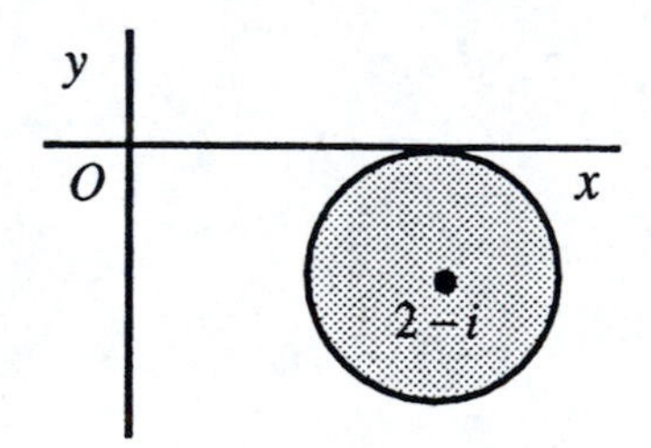

(b) Write $|2z+3| > 4$ as $\left|z-\left(-\frac{3}{2}\right)\right| > 2$ to see that the set in question consists of all points exterior to the circle with center at $-3/2$ and radius 2. It is a domain.

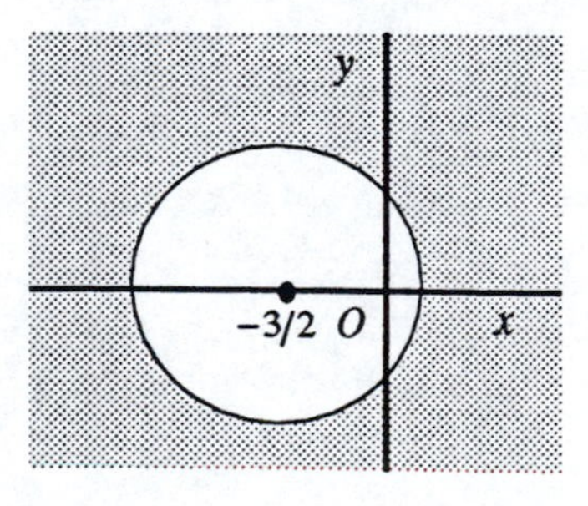

(c) Write $\operatorname{Im} z > 1$ as $y > 1$ to see that this is the half plane consisting of all points lying above the horizontal line $y = 1$. It is a domain.

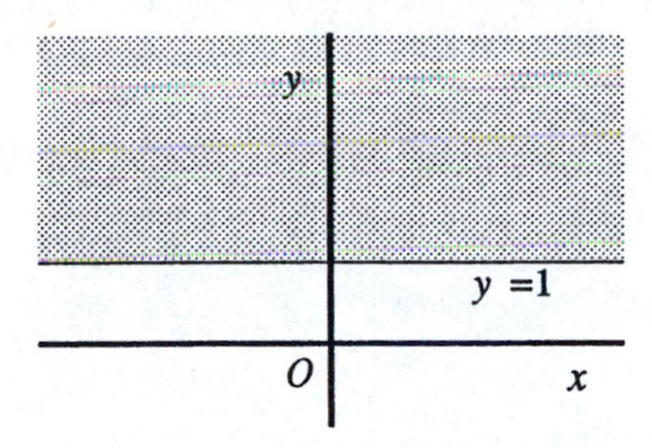

(d) The set $\operatorname{Im} z = 1$ is simply the horizontal line $y = 1$. It is *not* a domain.

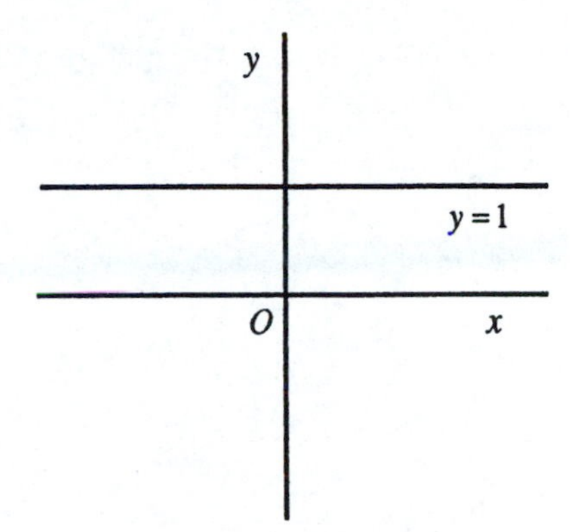

(e) The set $0 \le \arg z \le \dfrac{\pi}{4}$ $(z \ne 0)$ is indicated below. It is *not* a domain.

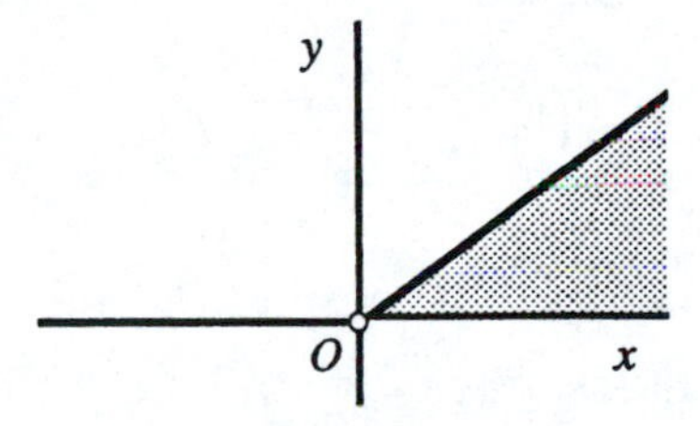

(f) The set $|z-4| \ge |z|$ can be written in the form $(x-4)^2 + y^2 \ge x^2 + y^2$, which reduces to $x \le 2$. This set, which is indicated below, is *not* a domain. The set is also geometrically evident since it consists of all points z such that the distance between z and 4 is greater than or equal to the distance between z and the origin.

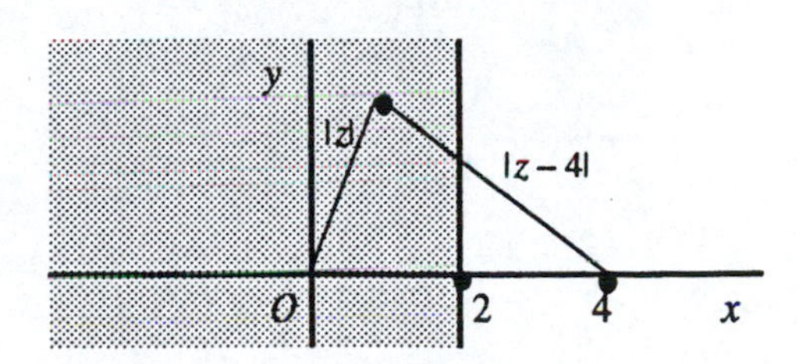

4. *(a)* The closure of the set $-\pi < \arg z < \pi \ (z \neq 0)$ is the entire plane.

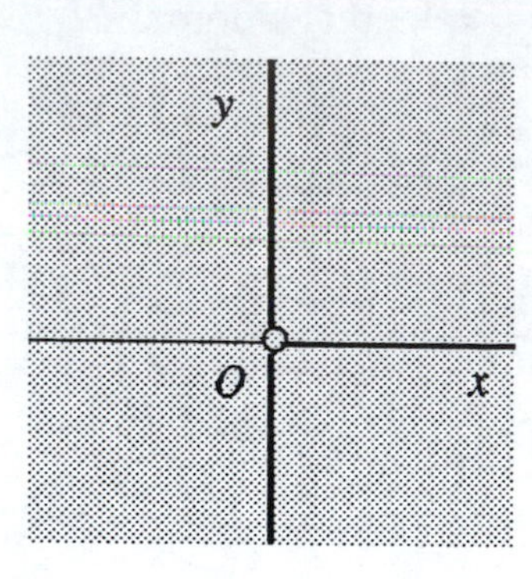

(b) We first write the set $|\text{Re}\, z| < |z|$ as $|x| < \sqrt{x^2 + y^2}$, or $x^2 < x^2 + y^2$. But this last inequality is the same as $y^2 > 0$, or $|y| > 0$. Hence the closure of the set $|\text{Re}\, z| < |z|$ is the entire plane.

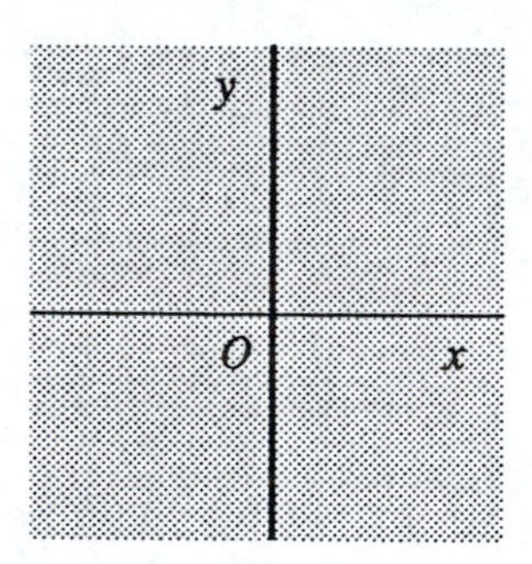

(c) Since $\dfrac{1}{z} = \dfrac{\bar{z}}{z\bar{z}} = \dfrac{\bar{z}}{|z|^2} = \dfrac{x - iy}{x^2 + y^2}$, the set $\text{Re}\left(\dfrac{1}{z}\right) \leq \dfrac{1}{2}$ can be written as $\dfrac{x}{x^2 + y^2} \leq \dfrac{1}{2}$, or $(x^2 - 2x) + y^2 \geq 0$. Finally, by completing the square, we arrive at the inequality $(x-1)^2 + y^2 \geq 1^2$, which describes the circle, together with its exterior, that is centered at $z = 1$ with radius 1. The closure of this set is itself.

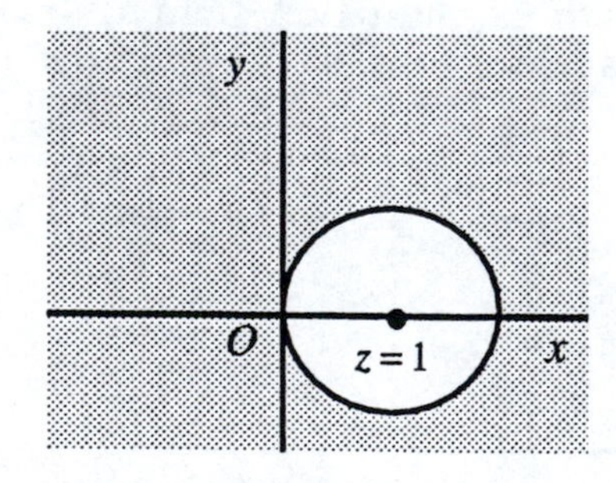

(d) Since $z^2 = (x+iy)^2 = x^2 - y^2 + i2xy$, the set $\text{Re}(z^2) > 0$ can be written as $y^2 < x^2$, or $|y| < |x|$. The closure of this set consists of the lines $y = \pm x$ together with the shaded region shown below.

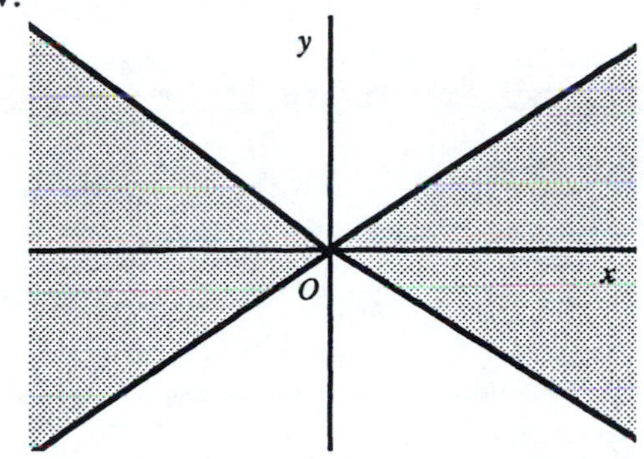

5. The set S consists of all points z such that $|z| < 1$ or $|z-2| < 1$, as shown below.

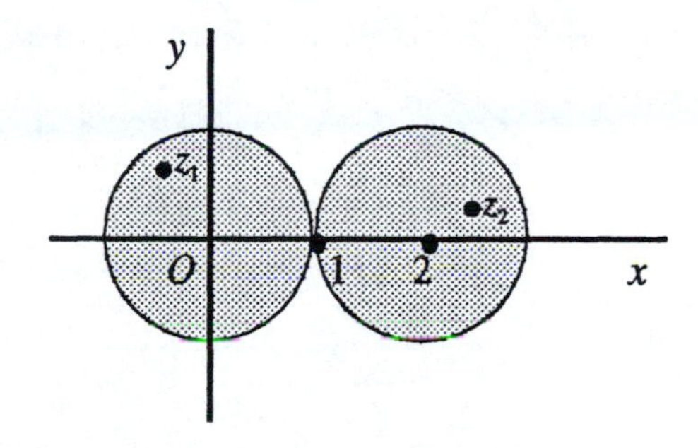

Since every polygonal line joining z_1 and z_2 must contain at least one point that is not in S, it is clear that S is not connected.

8. We are given that a set S contains each of its accumulation points. The problem here is to show that S must be closed. We do this by contradiction. We let z_0 be a boundary point of S and suppose that it is not a point in S. The fact that z_0 is a boundary point means that every neighborhood of z_0 contains at least one point in S; and, since z_0 is not in S, we see that every *deleted* neighborhood of S must contain at least one point in S. Thus z_0 is an accumulation point of S, and it follows that z_0 is a point in S. But this contradicts the fact that z_0 is not in S. We may conclude, then, that each boundary point z_0 must be in S. That is, S is closed.

Chapter 2

SECTION 11

1. *(a)* The function $f(z)=\dfrac{1}{z^2+1}$ is defined everywhere in the finite plane except at the points $z=\pm i$, where $z^2+1=0$.

(b) The function $f(z)=\operatorname{Arg}\left(\dfrac{1}{z}\right)$ is defined throughout the entire finite plane except for the point $z=0$.

(c) The function $f(z)=\dfrac{z}{z+\bar{z}}$ is defined everywhere in the finite plane except for the imaginary axis. This is because the equation $z+\bar{z}=0$ is the same as $x=0$.

(d) The function $f(z)=\dfrac{1}{1-|z|^2}$ is defined everywhere in the finite plane except on the circle $|z|=1$, where $1-|z|^2=0$.

3. Using $x=\dfrac{z+\bar{z}}{2}$ and $y=\dfrac{z-\bar{z}}{2i}$, write

$$\begin{aligned}
f(z) &= x^2-y^2-2y+i(2x-2xy)\\
&= \frac{(z+\bar{z})^2}{4}+\frac{(z-\bar{z})^2}{4}+i(z-\bar{z})+i(z+\bar{z})-\frac{(z+\bar{z})(z-\bar{z})}{2}\\
&= \frac{z^2}{2}+\frac{\bar{z}^2}{2}+2iz-\frac{z^2}{2}+\frac{\bar{z}^2}{2}=\bar{z}^2+2iz.
\end{aligned}$$

SECTION 17

5. Consider the function

$$f(z)=\left(\frac{z}{\bar{z}}\right)^2=\left(\frac{x+iy}{x-iy}\right)^2 \qquad (z\neq 0),$$

where $z=x+iy$. Observe that if $z=(x,0)$, then

$$f(z)=\left(\frac{x+i0}{x-i0}\right)^2=1;$$

and if $z=(0,y)$,

$$f(z)=\left(\frac{0+iy}{0-iy}\right)^2=1.$$

But if $z=(x,x)$,

$$f(z)=\left(\frac{x+ix}{x-ix}\right)^2=\left(\frac{1+i}{1-i}\right)^2=-1.$$

This shows that $f(z)$ has value 1 at all nonzero points on the real and imaginary axes but value -1 at all nonzero points on the line $y=x$. Thus the limit of $f(z)$ as z tends to 0 cannot exist.

10. *(a)* To show that $\lim\limits_{z\to\infty}\dfrac{4z^2}{(z-1)^2}=4$, we use statement (2), Sec. 16, and write

$$\lim_{z\to 0}\frac{4\left(\dfrac{1}{z}\right)^2}{\left(\dfrac{1}{z}-1\right)^2}=\lim_{z\to 0}\frac{4}{(1-z)^2}=4.$$

(b) To establish the limit $\lim\limits_{z\to 1}\dfrac{1}{(z-1)^3}=\infty$, we refer to statement (1), Sec. 16, and write

$$\lim_{z\to 1}\frac{1}{1/(z-1)^3}=\lim_{z\to 1}(z-1)^3=0.$$

(c) To verify that $\lim\limits_{z\to\infty}\dfrac{z^2+1}{z-1}=\infty$, we apply statement (3), Sec. 16, and write

$$\lim_{z\to 0}\frac{\dfrac{1}{z}-1}{\left(\dfrac{1}{z}\right)^2+1}=\lim_{z\to 0}\frac{z-z^2}{1+z^2}=0.$$

11. In this problem, we consider the function

$$T(z)=\frac{az+b}{cz+d}\qquad (ad-bc\neq 0).$$

(a) Suppose that $c=0$. Statement (3), Sec. 16, tells us that $\lim\limits_{z\to\infty}T(z)=\infty$ since

$$\lim_{z\to 0}\frac{1}{T(1/z)}=\lim_{z\to 0}\frac{c+dz}{a+bz}=\frac{c}{a}=0.$$

(b) Suppose that $c \neq 0$. Statement (2), Sec. 16, reveals that $\lim_{z\to\infty} T(z) = \frac{a}{c}$ since

$$\lim_{z\to 0} T\left(\frac{1}{z}\right) = \lim_{z\to 0} \frac{a+bz}{c+dz} = \frac{a}{c}.$$

Also, we know from statement (1), Sec. 16, that $\lim_{z\to -d/c} T(z) = \infty$ since

$$\lim_{z\to -d/c} \frac{1}{T(z)} = \lim_{z\to -d/c} \frac{cz+d}{az+b} = 0.$$

SECTION 19

1. *(a)* If $f(z) = 3z^2 - 2z + 4$, then

$$f'(z) = \frac{d}{dz}(3z^2 - 2z + 4) = 3\frac{d}{dz}z^2 - 2\frac{d}{dz}z + \frac{d}{dz}4 = 3(2z) - 2(1) + 0 = 6z - 2.$$

(b) If $f(z) = (1 - 4z^2)^3$, then

$$f'(z) = 3(1-4z^2)^2 \frac{d}{dz}(1-4z^2) = 3(1-4z^2)^2(-8z) = -24z(1-4z^2)^2.$$

(c) If $f(z) = \dfrac{z-1}{2z+1} \quad \left(z \neq -\dfrac{1}{2}\right)$, then

$$f'(z) = \frac{(2z+1)\frac{d}{dz}(z-1) - (z-1)\frac{d}{dz}(2z+1)}{(2z+1)^2} = \frac{(2z+1)(1) - (z-1)2}{(2z+1)^2} = \frac{3}{(2z+1)^2}.$$

(d) If $f(z) = \dfrac{(1+z^2)^4}{z^2}$ $(z \neq 0)$, then

$$f'(z) = \frac{z^2\frac{d}{dz}(1+z^2)^4 - (1+z^2)^4\frac{d}{dz}z^2}{(z^2)^2} = \frac{z^2 4(1+z^2)^3(2z) - (1+z^2)^4 2z}{(z^2)^2}$$

$$= \frac{2z(1+z^2)^3[4z^2 - (1+z^2)]}{z^4} = \frac{2(1+z^2)^3(3z^2-1)}{z^3}.$$

3. If $f(z)=1/z$ $(z\neq 0)$, then

$$\Delta w = f(z+\Delta z)-f(z)=\frac{1}{z+\Delta z}-\frac{1}{z}=\frac{-\Delta z}{(z+\Delta z)z}.$$

Hence

$$f'(z)=\lim_{\Delta z\to 0}\frac{\Delta w}{\Delta z}=\lim_{\Delta z\to 0}\frac{-1}{(z+\Delta z)z}=-\frac{1}{z^2}.$$

4. We are given that $f(z_0)=g(z_0)=0$ and that $f'(z_0)$ and $g'(z_0)$ exist, where $g'(z_0)\neq 0$. According to the definition of derivative,

$$f'(z_0)=\lim_{z\to z_0}\frac{f(z)-f(z_0)}{z-z_0}=\lim_{z\to z_0}\frac{f(z)}{z-z_0}.$$

Similarly,

$$g'(z_0)=\lim_{z\to z_0}\frac{g(z)-g(z_0)}{z-z_0}=\lim_{z\to z_0}\frac{g(z)}{z-z_0}.$$

Thus

$$\lim_{z\to z_0}\frac{f(z)}{g(z)}=\lim_{z\to z_0}\frac{f(z)/(z-z_0)}{g(z)/(z-z_0)}=\frac{\lim\limits_{z\to z_0} f(z)/(z-z_0)}{\lim\limits_{z\to z_0} g(z)/(z-z_0)}=\frac{f'(z_0)}{g'(z_0)}.$$

SECTION 22

1. *(a)* $f(z)=\bar{z}=x-iy$. So $u=x,\ v=-y$.
Inasmuch as $u_x=v_y \Rightarrow 1=-1$, the Cauchy-Riemann equations are not satisfied anywhere.

(b) $f(z)=z-\bar{z}=(x+iy)-(x-iy)=0+i2y$. So $u=0,\ v=2y$.
Since $u_x=v_y\Rightarrow 0=2$, the Cauchy-Riemann equations are not satisfied anywhere.

(c) $f(z)=2x+ixy^2$. Here $u=2x,\ v=xy^2$.
$u_x=v_y\Rightarrow 2=2xy\Rightarrow xy=1.$
$u_y=-v_x\Rightarrow 0=-y^2\Rightarrow y=0.$
Substituting $y=0$ into $xy=1$, we have $0=1$. Thus the Cauchy-Riemann equations do not hold anywhere.

(d) $f(z)=e^xe^{-iy}=e^x(\cos y-i\sin y)=e^x\cos y-ie^x\sin y$. So $u=e^x\cos y,\ v=-e^x\sin y$.
$u_x=v_y\Rightarrow e^x\cos y=-e^x\cos y\Rightarrow 2e^x\cos y=0\Rightarrow \cos y=0$. Thus

$$y=\frac{\pi}{2}+n\pi \qquad (n=0,\pm1,\pm2,\ldots).$$

$u_y=-v_x\Rightarrow -e^x\sin y=e^x\sin y\Rightarrow 2e^x\sin y=0\Rightarrow \sin y=0$. Hence

$$y=n\pi \qquad (n=0,\pm1,\pm2,\ldots).$$

Since these are two different sets of values of y, the Cauchy-Riemann equations cannot be satisfied anywhere.

3. *(a)* $f(z)=\frac{1}{z}=\frac{1}{z}\cdot\frac{\bar{z}}{\bar{z}}=\frac{\bar{z}}{|z|^2}=\frac{x}{x^2+y^2}+i\frac{-y}{x^2+y^2}$. So

$$u=\frac{x}{x^2+y^2}\quad\text{and}\quad v=\frac{-y}{x^2+y^2}.$$

Since

$$u_x=\frac{y^2-x^2}{(x^2+y^2)^2}=v_y\quad\text{and}\quad u_y=\frac{-2xy}{(x^2+y^2)^2}=-v_x\qquad(x^2+y^2\neq 0),$$

$f'(z)$ exists when $z\neq 0$. Moreover, when $z\neq 0$,

$$f'(z)=u_x+iv_x=\frac{y^2-x^2}{(x^2+y^2)^2}+i\frac{2xy}{(x^2+y^2)^2}=-\frac{x^2-i2xy-y^2}{(x^2+y^2)^2}$$

$$=-\frac{(x-iy)^2}{(x^2+y^2)^2}=-\frac{(\bar{z})^2}{(z\bar{z})^2}=-\frac{(\bar{z})^2}{(z)^2(\bar{z})^2}=-\frac{1}{z^2}.$$

(b) $f(z)=x^2+iy^2$. Hence $u=x^2$ and $v=y^2$. Now

$$u_x=v_y\Rightarrow 2x=2y\Rightarrow y=x\quad\text{and}\quad u_y=-v_x\Rightarrow 0=0.$$

So $f'(z)$ exists only when $y=x$, and we find that

$$f'(x+ix)=u_x(x,x)+iv_x(x,x)=2x+i0=2x.$$

(c) $f(z)=z\operatorname{Im}z=(x+iy)y=xy+iy^2$. Here $u=xy$ and $v=y^2$. We observe that

$$u_x=v_y\Rightarrow y=2y\Rightarrow y=0\quad\text{and}\quad u_y=-v_x\Rightarrow x=0.$$

Hence $f'(z)$ exists only when $z=0$. In fact,

$$f'(0)=u_x(0,0)+iv_x(0,0)=0+i0=0.$$

4. *(a)* $f(z)=\frac{1}{z^4}=\underbrace{\left(\frac{1}{r^4}\cos 4\theta\right)}_{u}+i\underbrace{\left(-\frac{1}{r^4}\sin 4\theta\right)}_{v}\quad(z\neq 0)$. Since

$$ru_r=-\frac{4}{r^4}\cos 4\theta=v_\theta\quad\text{and}\quad u_\theta=-\frac{4}{r^4}\sin 4\theta=-rv_r,$$

f is analytic in its domain of definition. Furthermore,

$$f'(z) = e^{-i\theta}(u_r + iv_r) = e^{-i\theta}\left(-\frac{4}{r^5}\cos 4\theta + i\frac{4}{r^5}\sin 4\theta\right)$$
$$= -\frac{4}{r^5}e^{-i\theta}(\cos 4\theta - i\sin 4\theta) = -\frac{4}{r^5}e^{-i\theta}e^{-i4\theta}$$
$$= \frac{-4}{r^5 e^{i5\theta}} = -\frac{4}{(re^{i\theta})^5} = -\frac{4}{z^5}.$$

(b) $f(z) = \sqrt{r}e^{i\theta/2} = \underbrace{\sqrt{r}\cos\frac{\theta}{2}}_{u} + i\underbrace{\sqrt{r}\sin\frac{\theta}{2}}_{v} \quad (r > 0, \alpha < \theta < \alpha + 2\pi)$. Since

$$ru_r = \frac{\sqrt{r}}{2}\cos\frac{\theta}{2} = v_\theta \quad \text{and} \quad u_\theta = -\frac{\sqrt{r}}{2}\sin\frac{\theta}{2} = -rv_r,$$

f is analytic in its domain of definition. Moreover,

$$f'(z) = e^{-i\theta}(u_r + iv_r) = e^{-i\theta}\left(\frac{1}{2\sqrt{r}}\cos\frac{\theta}{2} + i\frac{1}{2\sqrt{r}}\sin\frac{\theta}{2}\right)$$
$$= \frac{1}{2\sqrt{r}}e^{-i\theta}\left(\cos\frac{\theta}{2} + i\sin\frac{\theta}{2}\right) = \frac{1}{2\sqrt{r}}e^{-i\theta}e^{i\theta/2}$$
$$= \frac{1}{2\sqrt{r}e^{i\theta/2}} = \frac{1}{2f(z)}.$$

(c) $f(z) = \underbrace{e^{-\theta}\cos(\ln r)}_{u} + i\underbrace{e^{-\theta}\sin(\ln r)}_{v} \quad (r > 0, 0 < \theta < 2\pi)$. Since

$$ru_r = -e^{-\theta}\sin(\ln r) = v_\theta \quad \text{and} \quad u_\theta = -e^{-\theta}\cos(\ln r) = -rv_r,$$

f is analytic in its domain of definition. Also,

$$f'(z) = e^{-i\theta}(u_r + iv_r) = e^{-i\theta}\left[-\frac{e^{-\theta}\sin(\ln r)}{r} + i\frac{e^{-\theta}\cos(\ln r)}{r}\right]$$
$$= \frac{i}{re^{i\theta}}\left[e^{-\theta}\cos(\ln r) + ie^{-\theta}\sin(\ln r)\right] = i\frac{f(z)}{z}.$$

5. When $f(z) = x^3 + i(1-y)^3$, we have $u = x^3$ and $v = (1-y)^3$. Observe that

$$u_x = v_y \Rightarrow 3x^2 = -3(1-y)^2 \Rightarrow x^2 + (1-y)^2 = 0 \quad \text{and} \quad u_y = -v_x \Rightarrow 0 = 0.$$

Evidently, then, the Cauchy-Riemann equations are satisfied only when $x=0$ and $y=1$. That is, they hold only when $z = i$. Hence the expression

$$f'(z) = u_x + iv_x = 3x^2 + i0 = 3x^2$$

is valid only when $z = i$, in which case we see that $f'(i) = 0$.

6. Here u and v denote the real and imaginary components of the function f defined by means of the equations

$$f(z) = \begin{cases} \dfrac{\bar{z}^2}{z} & \text{when } z \neq 0, \\ 0 & \text{when } z = 0. \end{cases}$$

Now

$$f(z) = \underbrace{\frac{x^3 - 3xy^2}{x^2 + y^2}}_{u} + i\underbrace{\frac{y^3 - 3x^2y}{x^2 + y^2}}_{v}$$

when $z \neq 0$, and the following calculations show that

$$u_x(0,0) = v_y(0,0) \quad \text{and} \quad u_y(0,0) = -v_x(0,0):$$

$$u_x(0,0) = \lim_{\Delta x \to 0} \frac{u(0+\Delta x,0) - u(0,0)}{\Delta x} = \lim_{\Delta x \to 0} \frac{\Delta x}{\Delta x} = 1,$$

$$u_y(0,0) = \lim_{\Delta y \to 0} \frac{u(0,0+\Delta y) - u(0,0)}{\Delta y} = \lim_{\Delta y \to 0} \frac{0}{\Delta y} = 0,$$

$$v_x(0,0) = \lim_{\Delta x \to 0} \frac{v(0+\Delta x,0) - v(0,0)}{\Delta x} = \lim_{\Delta x \to 0} \frac{0}{\Delta x} = 0,$$

$$v_y(0,0) = \lim_{\Delta y \to 0} \frac{v(0,0+\Delta y) - v(0,0)}{\Delta y} = \lim_{\Delta y \to 0} \frac{\Delta y}{\Delta y} = 1.$$

7. Equations (2), Sec. 22, are

$$u_x \cos\theta + u_y \sin\theta = u_r,$$
$$-u_x r \sin\theta + u_y r \cos\theta = u_\theta.$$

Solving these simultaneous linear equations for u_x and u_y, we find that

$$u_x = u_r \cos\theta - u_\theta \frac{\sin\theta}{r} \quad \text{and} \quad u_y = u_r \sin\theta + u_\theta \frac{\cos\theta}{r}.$$

Likewise,

$$v_x = v_r \cos\theta - v_\theta \frac{\sin\theta}{r} \quad \text{and} \quad v_y = v_r \sin\theta + v_\theta \frac{\cos\theta}{r}.$$

Assume now that the Cauchy-Riemann equations in polar form,

$$ru_r = v_\theta, \quad u_\theta = -rv_r,$$

are satisfied at z_0. It follows that

$$u_x = u_r \cos\theta - u_\theta \frac{\sin\theta}{r} = v_\theta \frac{\cos\theta}{r} + v_r \sin\theta = v_r \sin\theta + v_\theta \frac{\cos\theta}{r} = v_y,$$

$$u_y = u_r \sin\theta + u_\theta \frac{\cos\theta}{r} = v_\theta \frac{\sin\theta}{r} - v_r \cos\theta = -\left(v_r \cos\theta - v_\theta \frac{\sin\theta}{r} \right) = -v_x.$$

9. *(a)* Write $f(z) = u(r,\theta) + iv(r,\theta)$. Then recall the polar form

$$ru_r = v_\theta, \quad u_\theta = -rv_r$$

of the Cauchy-Riemann equations, which enables us to rewrite the expression (Sec. 22)

$$f'(z_0) = e^{-i\theta}(u_r + iv_r)$$

for the derivative of f at a point $z_0 = (r_0, \theta_0)$ in the following way:

$$f'(z_0) = e^{-i\theta}\left(\frac{1}{r} v_\theta - \frac{i}{r} u_\theta \right) = \frac{-i}{re^{i\theta}}(u_\theta + iv_\theta) = \frac{-i}{z_0}(u_\theta + iv_\theta).$$

(b) Consider now the function

$$f(z)=\frac{1}{z}=\frac{1}{re^{i\theta}}=\frac{1}{r}e^{-i\theta}=\frac{1}{r}(\cos\theta-i\sin\theta)=\frac{\cos\theta}{r}-i\frac{\sin\theta}{r}.$$

With

$$u(r,\theta)=\frac{\cos\theta}{r} \quad \text{and} \quad v(r,\theta)=-\frac{\sin\theta}{r},$$

the final expression for $f'(z_0)$ in part *(a)* tells us that

$$f'(z)=\frac{-i}{z}\left(-\frac{\sin\theta}{r}-i\frac{\cos\theta}{r}\right)=-\frac{1}{z}\left(\frac{\cos\theta-i\sin\theta}{r}\right)$$

$$=-\frac{1}{z}\left(\frac{e^{-i\theta}}{r}\right)=-\frac{1}{z}\left(\frac{1}{re^{i\theta}}\right)=-\frac{1}{z^2}$$

when $z\neq 0$.

10. *(a)* We consider a function $F(x,y)$, where

$$x=\frac{z+\bar{z}}{2} \quad \text{and} \quad y=\frac{z-\bar{z}}{2i}.$$

Formal application of the chain rule for multivariable functions yields

$$\frac{\partial F}{\partial \bar{z}}=\frac{\partial F}{\partial x}\frac{\partial x}{\partial \bar{z}}+\frac{\partial F}{\partial y}\frac{\partial y}{\partial \bar{z}}=\frac{\partial F}{\partial x}\left(\frac{1}{2}\right)+\frac{\partial F}{\partial y}\left(-\frac{1}{2i}\right)=\frac{1}{2}\left(\frac{\partial F}{\partial x}+i\frac{\partial F}{\partial y}\right).$$

(b) Now define the operator

$$\frac{\partial}{\partial \bar{z}}=\frac{1}{2}\left(\frac{\partial}{\partial x}+i\frac{\partial}{\partial y}\right),$$

suggested by part *(a)*, and formally apply it to a function $f(z)=u(x,y)+iv(x,y)$:

$$\frac{\partial f}{\partial \bar{z}}=\frac{1}{2}\left(\frac{\partial f}{\partial x}+i\frac{\partial f}{\partial y}\right)=\frac{1}{2}\frac{\partial f}{\partial x}+\frac{i}{2}\frac{\partial f}{\partial y}$$

$$=\frac{1}{2}(u_x+iv_x)+\frac{i}{2}(u_y+iv_y)=\frac{1}{2}\left[(u_x-v_y)+i(v_x+u_y)\right].$$

If the Cauchy-Riemann equations $u_x=v_y$, $u_y=-v_x$ are satisfied, this tells us that $\partial f/\partial \bar{z}=0$.

SECTION 24

1. *(a)* $f(z) = \underbrace{3x + y}_{u} + i\underbrace{(3y - x)}_{v}$ is entire since

$$u_x = 3 = v_y \quad \text{and} \quad u_y = 1 = -v_x.$$

(b) $f(z) = \underbrace{\sin x \cosh y}_{u} + i\underbrace{\cos x \sinh y}_{v}$ is entire since

$$u_x = \cos x \cosh y = v_y \quad \text{and} \quad u_y = \sin x \sinh y = -v_x.$$

(c) $f(z) = e^{-y}\sin x - ie^{-y}\cos x = \underbrace{e^{-y}\sin x}_{u} + i\underbrace{(-e^{-y}\cos x)}_{v}$ is entire since

$$u_x = e^{-y}\cos x = v_y \quad \text{and} \quad u_y = -e^{-y}\sin x = -v_x.$$

(d) $f(z) = (z^2 - 2)e^{-x}e^{-iy}$ is entire since it is the product of the entire functions

$$g(z) = z^2 - 2 \quad \text{and} \quad h(z) = e^{-x}e^{-iy} = e^{-x}(\cos y - i\sin y) = \underbrace{e^{-x}\cos y}_{u} + i\underbrace{(-e^{-x}\sin y)}_{v}.$$

The function g is entire since it is a polynomial, and h is entire since

$$u_x = -e^{-x}\cos y = v_y \quad \text{and} \quad u_y = -e^{-x}\sin y = -v_x.$$

2. *(a)* $f(z) = \underbrace{xy}_{u} + i\underbrace{y}_{v}$ is nowhere analytic since

$$u_x = v_y \Rightarrow y = 1 \quad \text{and} \quad u_y = -v_x \Rightarrow x = 0,$$

which means that the Cauchy-Riemann equations hold only at the point $z = (0,1) = i$.

(c) $f(z) = e^y e^{ix} = e^y(\cos x + i\sin x) = \underbrace{e^y\cos x}_{u} + i\underbrace{e^y\sin x}_{v}$ is nowhere analytic since

$$u_x = v_y \Rightarrow -e^y\sin x = e^y\sin x \Rightarrow 2e^y\sin x = 0 \Rightarrow \sin x = 0$$

and

$$u_y = -v_x \Rightarrow e^y\cos x = -e^y\cos x \Rightarrow 2e^y\cos x = 0 \Rightarrow \cos x = 0.$$

More precisely, the roots of the equation $\sin x = 0$ are $n\pi$ $(n = 0, \pm 1, \pm 2, \ldots)$, and $\cos n\pi = (-1)^n \neq 0$. Consequently, the Cauchy-Riemann equations are not satisfied anywhere.

7. *(a)* Suppose that a function $f(z) = u(x,y) + iv(x,y)$ is analytic and real-valued in a domain D. Since $f(z)$ is real-valued, it has the form $f(z) = u(x,y) + i0$. The Cauchy-Riemann equations $u_x = v_y, u_y = -v_x$ thus become $u_x = 0, u_y = 0$; and this means that $u(x,y) = a$, where a is a (real) constant. (See the proof of the theorem in Sec. 23.) Evidently, then, $f(z) = a$. That is, f is constant in D.

(b) Suppose that a function f is analytic in a domain D and that its modulus $|f(z)|$ is constant there. Write $|f(z)| = c$, where c is a (real) constant. If $c = 0$, we see that $f(z) = 0$ throughout D. If, on the other hand, $c \neq 0$, write $f(z)\overline{f(z)} = c^2$, or

$$\overline{f(z)} = \frac{c^2}{f(z)}.$$

Since $f(z)$ is analytic and never zero in D, the conjugate $\overline{f(z)}$ must be analytic in D. Example 3 in Sec. 24 then tells us that $f(z)$ must be constant in D.

SECTION 25

1. *(a)* It is straightforward to show that $u_{xx} + u_{yy} = 0$ when $u(x,y) = 2x(1-y)$. To find a harmonic conjugate $v(x,y)$, we start with $u_x(x,y) = 2 - 2y$. Now

$$u_x = v_y \Rightarrow v_y = 2 - 2y \Rightarrow v(x,y) = 2y - y^2 + \phi(x).$$

Then

$$u_y = -v_x \Rightarrow -2x = -\phi'(x) \Rightarrow \phi'(x) = 2x \Rightarrow \phi(x) = x^2 + c.$$

Consequently,

$$v(x,y) = 2y - y^2 + (x^2 + c) = x^2 - y^2 + 2y + c.$$

(b) It is straightforward to show that $u_{xx} + u_{yy} = 0$ when $u(x,y) = 2x - x^3 + 3xy^2$. To find a harmonic conjugate $v(x,y)$, we start with $u_x(x,y) = 2 - 3x^2 + 3y^2$. Now

$$u_x = v_y \Rightarrow v_y = 2 - 3x^2 + 3y^2 \Rightarrow v(x,y) = 2y - 3x^2y + y^3 + \phi(x).$$

Then

$$u_y = -v_x \Rightarrow 6xy = 6xy - \phi'(x) \Rightarrow \phi'(x) = 0 \Rightarrow \phi(x) = c.$$

Consequently,

$$v(x,y) = 2y - 3x^2y + y^3 + c.$$

(c) It is straightforward to show that $u_{xx} + u_{yy} = 0$ when $u(x,y) = \sinh x \sin y$. To find a harmonic conjugate $v(x,y)$, we start with $u_x(x,y) = \cosh x \sin y$. Now

$$u_x = v_y \Rightarrow v_y = \cosh x \sin y \Rightarrow v(x,y) = -\cosh x \cos y + \phi(x).$$

Then

$$u_y = -v_x \Rightarrow \sinh x \cos y = \sinh x \cos y - \phi'(x) \Rightarrow \phi'(x) = 0 \Rightarrow \phi(x) = c.$$

Consequently,

$$v(x,y) = -\cosh x \cos y + c.$$

(d) It is straightforward to show that $u_{xx}+u_{yy}=0$ when $u(x,y)=\dfrac{y}{x^2+y^2}$. To find a harmonic conjugate $v(x,y)$, we start with $u_x(x,y)=-\dfrac{2xy}{(x^2+y^2)^2}$. Now

$$u_x=v_y \Rightarrow v_y=-\frac{2xy}{(x^2+y^2)^2} \Rightarrow v(x,y)=\frac{x}{x^2+y^2}+\phi(x).$$

Then

$$u_y=-v_x \Rightarrow \frac{x^2-y^2}{(x^2+y^2)^2}=\frac{x^2-y^2}{(x^2+y^2)^2}-\phi'(x) \Rightarrow \phi'(x)=0 \Rightarrow \phi(x)=c.$$

Consequently,

$$v(x,y)=\frac{x}{x^2+y^2}+c.$$

2. Suppose that v and V are harmonic conjugates of u in a domain D. This means that

$$u_x=v_y, \quad u_y=-v_x \quad \text{and} \quad u_x=V_y, \quad u_y=-V_x.$$

If $w=v-V$, then,

$$w_x=v_x-V_x=-u_y+u_y=0 \quad \text{and} \quad w_y=v_y-V_y=u_x-u_x=0.$$

Hence $w(x,y)=c$, where c is a (real) constant (compare the proof of the theorem in Sec. 23). That is, $v(x,y)-V(x,y)=c$.

3. Suppose that u and v are harmonic conjugates of each other in a domain D. Then

$$u_x=v_y, \quad u_y=-v_x \quad \text{and} \quad v_x=u_y, \quad v_y=-u_x.$$

It follows readily from these equations that

$$u_x=0, \quad u_y=0 \quad \text{and} \quad v_x=0, \quad v_y=0.$$

Consequently, $u(x,y)$ and $v(x,y)$ must be constant throughout D (compare the proof of the theorem in Sec. 23).

5. The Cauchy-Riemann equations in polar coordinates are

$$ru_r=v_\theta \quad \text{and} \quad u_\theta=-rv_r.$$

Now

$$ru_r=v_\theta \Rightarrow ru_{rr}+u_r=v_{\theta r}$$

and

$$u_\theta = -rv_r \Rightarrow u_{\theta\theta} = -rv_{r\theta}.$$

Thus

$$r^2u_{rr} + ru_r + u_{\theta\theta} = rv_{\theta r} - rv_{r\theta};$$

and, since $v_{\theta r} = v_{r\theta}$, we have

$$r^2u_{rr} + ru_r + u_{\theta\theta} = 0,$$

which is the polar form of Laplace's equation. To show that v satisfies the same equation, we observe that

$$u_\theta = -rv_r \Rightarrow v_r = -\frac{1}{r}u_\theta \Rightarrow v_{rr} = \frac{1}{r^2}u_\theta - \frac{1}{r}u_{\theta r}$$

and

$$ru_r = v_\theta \Rightarrow v_{\theta\theta} = ru_{r\theta}.$$

Since $u_{\theta r} = u_{r\theta}$, then,

$$r^2v_{rr} + rv_r + v_{\theta\theta} = u_\theta - ru_{\theta r} - u_\theta + ru_{r\theta} = 0.$$

6. If $u(r,\theta) = \ln r$, then

$$r^2u_{rr} + ru_r + u_{\theta\theta} = r^2\left(-\frac{1}{r^2}\right) + r\left(\frac{1}{r}\right) + 0 = 0.$$

This tells us that the function $u = \ln r$ is harmonic in the domain $r > 0, 0 < \theta < 2\pi$. Now it follows from the Cauchy-Riemann equation $ru_r = v_\theta$ and the derivative $u_r = \frac{1}{r}$ that $v_\theta = 1$; thus $v(r,\theta) = \theta + \phi(r)$, where $\phi(r)$ is at present an arbitrary differentiable function of r. The other Cauchy-Riemann equation $u_\theta = -rv_r$ then becomes $0 = -r\phi'(r)$. That is, $\phi'(r) = 0$; and we see that $\phi(r) = c$, where c is an arbitrary (real) constant. Hence $v(r,\theta) = \theta + c$ is a harmonic conjugate of $u(r,\theta) = \ln r$.

Chapter 3

SECTION 28

1. *(a)* $\exp(2\pm 3\pi i) = e^2 \exp(\pm 3\pi i) = -e^2$, since $\exp(\pm 3\pi i) = -1$.

(b) $$\exp\frac{2+\pi i}{4} = \left(\exp\frac{1}{2}\right)\left(\exp\frac{\pi i}{4}\right) = \sqrt{e}\left(\cos\frac{\pi}{4} + i\sin\frac{\pi}{4}\right)$$
$$= \sqrt{e}\left(\frac{1}{\sqrt{2}} + i\frac{1}{\sqrt{2}}\right) = \sqrt{\frac{e}{2}}(1+i).$$

(c) $\exp(z+\pi i) = (\exp z)(\exp \pi i) = -\exp z$, since $\exp \pi i = -1$.

3. First write
$$\exp(\bar{z}) = \exp(x - iy) = e^x e^{-iy} = e^x \cos y - ie^x \sin y,$$
where $z = x + iy$. This tells us that $\exp(\bar{z}) = u(x,y) + iv(x,y)$, where
$$u(x,y) = e^x \cos y \quad \text{and} \quad v(x,y) = -e^x \sin y.$$
Suppose that the Cauchy-Riemann equations $u_x = v_y$ and $u_y = -v_x$ are satisfied at some point $z = x + iy$. It is easy to see that, for the functions u and v here, these equations become $\cos y = 0$ and $\sin y = 0$. But there is no value of y satisfying this pair of equations. We may conclude that, since the Cauchy-Riemann equations fail to be satisfied anywhere, the function $\exp(\bar{z})$ is not analytic anywhere.

4. The function $\exp(z^2)$ is entire since it is a composition of the entire functions z^2 and $\exp z$; and the chain rule for derivatives tells us that
$$\frac{d}{dz}\exp(z^2) = \exp(z^2)\frac{d}{dz}z^2 = 2z\exp(z^2).$$
Alternatively, one can show that $\exp(z^2)$ is entire by writing
$$\exp(z^2) = \exp\left[(x+iy)^2\right] = \exp(x^2 - y^2)\exp(i2xy)$$
$$= \underbrace{\exp(x^2 - y^2)\cos(2xy)}_{u} + i\underbrace{\exp(x^2 - y^2)\sin(2xy)}_{v}$$
and using the Cauchy-Riemann equations. To be specific,
$$u_x = 2x\exp(x^2 - y^2)\cos(2xy) - 2y\exp(x^2 - y^2)\sin(2xy) = v_y$$
and
$$u_y = -2y\exp(x^2 - y^2)\cos(2xy) - 2x\exp(x^2 - y^2)\sin(2xy) = -v_x.$$

Furthermore,

$$\frac{d}{dz}\exp(z^2) = u_x + iv_x = 2(x+iy)\left[\exp(x^2-y^2)\cos(2xy) + i\exp(x^2-y^2)\sin(2xy)\right]$$
$$= 2z\exp(z^2).$$

5. We first write

$$|\exp(2z+i)| = |\exp[2x+i(2y+1)]| = e^{2x}$$

and

$$|\exp(iz^2)| = |\exp[-2xy+i(x^2-y^2)]| = e^{-2xy}.$$

Then, since

$$|\exp(2z+i)+\exp(iz^2)| \le |\exp(2z+i)| + |\exp(iz^2)|,$$

it follows that

$$|\exp(2z+i)+\exp(iz^2)| \le e^{2x} + e^{-2xy}.$$

6. First write

$$|\exp(z^2)| = |\exp[(x+iy)^2]| = |\exp(x^2-y^2)+i2xy| = \exp(x^2-y^2)$$

and

$$\exp(|z|^2) = \exp(x^2+y^2).$$

Since $x^2-y^2 \le x^2+y^2$, it is clear that $\exp(x^2-y^2) \le \exp(x^2+y^2)$. Hence it follows from the above that

$$|\exp(z^2)| \le \exp(|z|^2).$$

7. To prove that $|\exp(-2z)| < 1 \Leftrightarrow \operatorname{Re} z > 0$, write

$$|\exp(-2z)| = |\exp(-2x-i2y)| = \exp(-2x).$$

It is then clear that the statement to be proved is the same as $\exp(-2x) < 1 \Leftrightarrow x > 0$, which is obvious from the graph of the exponential function in calculus.

8. *(a)* Write $e^z=-2$ as $e^x e^{iy}=2e^{i\pi}$. This tells us that

$$e^x=2 \quad \text{and} \quad y=\pi+2n\pi \qquad (n=0,\pm1,\pm2,\ldots).$$

That is,

$$x=\ln 2 \quad \text{and} \quad y=(2n+1)\pi \qquad (n=0,\pm1,\pm2,\ldots).$$

Hence

$$z=\ln 2+(2n+1)\pi i \qquad (n=0,\pm1,\pm2,\ldots).$$

(b) Write $e^z=1+\sqrt{3}i$ as $e^x e^{iy}=2e^{i(\pi/3)}$, from which we see that

$$e^x=2 \quad \text{and} \quad y=\frac{\pi}{3}+2n\pi \qquad (n=0,\pm1,\pm2,\ldots).$$

That is,

$$x=\ln 2 \quad \text{and} \quad y=\left(2n+\frac{1}{3}\right)\pi \qquad (n=0,\pm1,\pm2,\ldots).$$

Consequently,

$$z=\ln 2+\left(2n+\frac{1}{3}\right)\pi i \qquad (n=0,\pm1,\pm2,\ldots).$$

(c) Write $\exp(2z-1)=1$ as $e^{2x-1}e^{i2y}=1e^{i0}$ and note how it follows that

$$e^{2x-1}=1 \quad \text{and} \quad 2y=0+2n\pi \qquad (n=0,\pm1,\pm2,\ldots).$$

Evidently, then,

$$x=\frac{1}{2} \quad \text{and} \quad y=n\pi \qquad (n=0,\pm1,\pm2,\ldots);$$

and this means that

$$z=\frac{1}{2}+n\pi i \qquad (n=0,\pm1,\pm2,\ldots).$$

9. This problem is actually to find all roots of the equation

$$\overline{\exp(iz)}=\exp(i\bar{z}).$$

To do this, set $z = x + iy$ and rewrite the equation as

$$e^{-y}e^{-ix} = e^{y}e^{ix}.$$

Now, according to the statement in italics at the beginning of Sec.8 in the text,

$$e^{-y} = e^{y} \quad \text{and} \quad -x = x + 2n\pi,$$

where n may have any one of the values $n = 0, \pm 1, \pm 2, \ldots$. Thus

$$y = 0 \quad \text{and} \quad x = n\pi \qquad (n = 0, \pm 1, \pm 2, \ldots).$$

The roots of the original equation are, therefore,

$$z = n\pi \qquad (n = 0, \pm 1, \pm 2, \ldots).$$

10. *(a)* Suppose that e^z is real. Since $e^z = e^x \cos y + ie^x \sin y$, this means that $e^x \sin y = 0$. Moreover, since e^x is never zero, $\sin y = 0$. Consequently, $y = n\pi\ (n = 0, \pm 1, \pm 2, \ldots)$; that is, $\operatorname{Im} z = n\pi\ (n = 0, \pm 1, \pm 2, \ldots)$.

(b) On the other hand, suppose that e^z is pure imaginary. It follows that $\cos y = 0$, or that $y = \dfrac{\pi}{2} + n\pi\ (n = 0, \pm 1, \pm 2, \ldots)$. That is, $\operatorname{Im} z = \dfrac{\pi}{2} + n\pi\ (n = 0, \pm 1, \pm 2, \ldots)$.

12. We start by writing

$$\frac{1}{z} = \frac{\bar{z}}{z\bar{z}} = \frac{\bar{z}}{|z|^2} = \frac{x - iy}{x^2 + y^2} = \frac{x}{x^2 + y^2} + i\frac{-y}{x^2 + y^2}.$$

Because $\operatorname{Re}(e^z) = e^x \cos y$, it follows that

$$\operatorname{Re}(e^{1/z}) = \exp\left(\frac{x}{x^2 + y^2}\right)\cos\left(\frac{-y}{x^2 + y^2}\right) = \exp\left(\frac{x}{x^2 + y^2}\right)\cos\left(\frac{y}{x^2 + y^2}\right).$$

Since $e^{1/z}$ is analytic in every domain that does not contain the origin, Theorem 1 in Sec. 25 ensures that $\operatorname{Re}(e^{1/z})$ is harmonic in such a domain.

13. If $f(z) = u(x, y) + iv(x, y)$ is analytic in some domain D, then

$$e^{f(z)} = e^{u(x,y)} \cos v(x, y) + ie^{u(x,y)} \sin v(x, y).$$

Since $e^{f(z)}$ is a composition of functions that are analytic in D, it follows from Theorem 1 in Sec. 25 that its component functions

$$U(x, y) = e^{u(x,y)} \cos v(x, y), \quad V(x, y) = e^{u(x,y)} \sin v(x, y)$$

are harmonic in D. Moreover, by Theorem 2 in Sec. 25, $V(x,y)$ is a harmonic conjugate of $U(x,y)$.

14. The problem here is to establish the identity

$$(\exp z)^n = \exp(nz) \qquad (n = 0, \pm 1, \pm 2, \ldots).$$

(a) To show that it is true when $n = 0,1,2,\ldots$, we use mathematical induction. It is obviously true when $n = 0$. Suppose that it is true when $n = m$, where m is any nonnegative integer. Then

$$(\exp z)^{m+1} = (\exp z)^m (\exp z) = \exp(mz)\exp z = \exp(mz + z) = \exp[(m+1)z].$$

(b) Suppose now that n is a negative integer $(n = -1, -2, \ldots)$, and write $m = -n = 1, 2, \ldots$. In view of part *(a)*,

$$(\exp z)^n = \left(\frac{1}{\exp z}\right)^m = \frac{1}{(\exp z)^m} = \frac{1}{\exp(mz)} = \frac{1}{\exp(-nz)} = \exp(nz).$$

SECTION 30

1. *(a)* $\text{Log}(-ei) = \ln|-ei| + i\text{Arg}(-ei) = \ln e - \dfrac{\pi}{2}i = 1 - \dfrac{\pi}{2}i.$

(b) $\text{Log}(1-i) = \ln|1-i| + i\text{Arg}(1-i) = \ln\sqrt{2} - \dfrac{\pi}{4}i = \dfrac{1}{2}\ln 2 - \dfrac{\pi}{4}i.$

2. *(a)* $\log e = \ln e + i(0 + 2n\pi) = 1 + 2n\pi i \quad (n = 0, \pm 1, \pm 2, \ldots).$

(b) $\log i = \ln 1 + i\left(\dfrac{\pi}{2} + 2n\pi\right) = \left(2n + \dfrac{1}{2}\right)\pi i \quad (n = 0, \pm 1, \pm 2, \ldots).$

(c) $\log\left(-1 + \sqrt{3}i\right) = \ln 2 + i\left(\dfrac{2\pi}{3} + 2n\pi\right) = \ln 2 + 2\left(n + \dfrac{1}{3}\right)\pi i \quad (n = 0, \pm 1, \pm 2, \ldots).$

3. *(a)* Observe that

$$\text{Log}(1+i)^2 = \text{Log}(2i) = \ln 2 + \frac{\pi}{2}i$$

and

$$2\text{Log}(1+i) = 2\left(\ln\sqrt{2} + i\frac{\pi}{4}\right) = \ln 2 + \frac{\pi}{2}i.$$

Thus

$$\text{Log}(1+i)^2 = 2\text{Log}(1+i).$$

(b) On the other hand,

$$\mathrm{Log}(-1+i)^2 = \mathrm{Log}(-2i) = \ln 2 - \frac{\pi}{2}i$$

and

$$2\mathrm{Log}(-1+i) = 2\left(\ln\sqrt{2} + i\frac{3\pi}{4}\right) = \ln 2 + \frac{3\pi}{2}i.$$

Hence

$$\mathrm{Log}(-1+i)^2 \neq 2\mathrm{Log}(-1+i).$$

4. *(a)* Consider the branch

$$\log z = \ln r + i\theta \qquad \left(r > 0, \frac{\pi}{4} < \theta < \frac{9\pi}{4}\right).$$

Since

$$\log(i^2) = \log(-1) = \ln 1 + i\pi = \pi i \quad \text{and} \quad 2\log i = 2\left(\ln 1 + i\frac{\pi}{2}\right) = \pi i,$$

we find that $\log(i^2) = 2\log i$ when this branch of $\log z$ is taken.

(b) Now consider the branch

$$\log z = \ln r + i\theta \qquad \left(r > 0, \frac{3\pi}{4} < \theta < \frac{11\pi}{4}\right).$$

Here

$$\log(i^2) = \log(-1) = \ln 1 + i\pi = \pi i \quad \text{and} \quad 2\log i = 2\left(\ln 1 + i\frac{5\pi}{2}\right) = 5\pi i.$$

Hence, for this particular branch, $\log(i^2) \neq 2\log i$.

5. *(a)* The two values of $i^{1/2}$ are $e^{i\pi/4}$ and $e^{i5\pi/4}$. Observe that

$$\log(e^{i\pi/4}) = \ln 1 + i\left(\frac{\pi}{4} + 2n\pi\right) = \left(2n + \frac{1}{4}\right)\pi i \qquad (n = 0, \pm 1, \pm 2, \ldots)$$

and

$$\log(e^{i5\pi/4}) = \ln 1 + i\left(\frac{5\pi}{4} + 2n\pi\right) = \left[(2n+1) + \frac{1}{4}\right]\pi i \qquad (n = 0, \pm 1, \pm 2, \ldots).$$

Combining these two sets of values, we find that

$$\log(i^{1/2}) = \left(n + \frac{1}{4}\right)\pi i \qquad (n = 0, \pm 1, \pm 2, \ldots).$$

On the other hand,

$$\frac{1}{2}\log i = \frac{1}{2}\left[\ln 1 + i\left(\frac{\pi}{2} + 2n\pi\right)\right] = \left(n + \frac{1}{4}\right)\pi i \qquad (n = 0, \pm 1, \pm 2, \ldots).$$

Thus the set of values of $\log(i^{1/2})$ is the same as the set of values of $\frac{1}{2}\log i$, and we may write

$$\log(i^{1/2}) = \frac{1}{2}\log i.$$

(b) Note that

$$\log(i^2) = \log(-1) = \ln 1 + (\pi + 2n\pi)i = (2n+1)\pi i \qquad (n = 0, \pm 1, \pm 2, \ldots)$$

but that

$$2\log i = 2\left[\ln 1 + i\left(\frac{\pi}{2} + 2n\pi\right)\right] = (4n+1)\pi i \qquad (n = 0, \pm 1, \pm 2, \ldots).$$

Evidently, then, the set of values of $\log(i^2)$ is *not* the same as the set of values of $2\log i$. That is,

$$\log(i^2) \neq 2\log i.$$

7. To solve the equation $\log z = i\pi/2$, write $\exp(\log z) = \exp(i\pi/2)$, or $z = e^{i\pi/2} = i$.

10. Since $\ln(x^2 + y^2)$ is the real component of any (analytic) branch of $2\log z$, it is harmonic in every domain that does not contain the origin. This can be verified directly by writing $u(x,y) = \ln(x^2 + y^2)$ and showing that $u_{xx}(x,y) + u_{yy}(x,y) = 0$.

SECTION 31

1. Suppose that $\text{Re}\, z_1 > 0$ and $\text{Re}\, z_2 > 0$. Then

$$z_1 = r_1 \exp i\Theta_1 \quad \text{and} \quad z_2 = r_2 \exp i\Theta_2,$$

where

$$-\frac{\pi}{2} < \Theta_1 < \frac{\pi}{2} \quad \text{and} \quad -\frac{\pi}{2} < \Theta_2 < \frac{\pi}{2}.$$

The fact that $-\pi < \Theta_1 + \Theta_2 < \pi$ enables us to write

$$\operatorname{Log}(z_1 z_2) = \operatorname{Log}[(r_1 r_2)\exp i(\Theta_1 + \Theta_2)] = \ln(r_1 r_2) + i(\Theta_1 + \Theta_2)$$

$$= (\ln r_1 + i\Theta_1) + (\ln r_2 + i\Theta_2) = \operatorname{Log}(r_1 \exp i\Theta_1) + \operatorname{Log}(r_2 \exp i\Theta_2)$$

$$= \operatorname{Log} z_1 + \operatorname{Log} z_2.$$

3. We are asked to show in two different ways that

$$\log\left(\frac{z_1}{z_2}\right) = \log z_1 - \log z_2 \qquad (z_1 \neq 0, z_2 \neq 0).$$

(a) One way is to refer to the relation $\arg\left(\frac{z_1}{z_2}\right) = \arg z_1 - \arg z_2$ in Sec. 7 and write

$$\log\left(\frac{z_1}{z_2}\right) = \ln\left|\frac{z_1}{z_2}\right| + i\arg\left(\frac{z_1}{z_2}\right) = (\ln|z_1| + i\arg z_1) - (\ln|z_2| + i\arg z_2) = \log z_1 - \log z_2.$$

(b) Another way is to first show that $\log\left(\frac{1}{z}\right) = -\log z \; (z \neq 0)$. To do this, we write $z = re^{i\theta}$

and then

$$\log\left(\frac{1}{z}\right) = \log\left(\frac{1}{r}e^{-i\theta}\right) = \ln\left(\frac{1}{r}\right) + i(-\theta + 2n\pi) = -[\ln r + i(\theta - 2n\pi)] = -\log z,$$

where $n = 0, \pm 1, \pm 2, \ldots$. This enables us to use the relation

$$\log(z_1 z_2) = \log z_1 + \log z_2$$

and write

$$\log\left(\frac{z_1}{z_2}\right) = \log\left(z_1 \frac{1}{z_2}\right) = \log z_1 + \log\left(\frac{1}{z_2}\right) = \log z_1 - \log z_2.$$

5. The problem here is to verify that

$$z^{1/n} = \exp\left(\frac{1}{n}\log z\right) \qquad (n = -1,-2,\ldots),$$

given that it is valid when $n = 1,2,\ldots$. To do this, we put $m = -n$, where n is a negative integer. Then, since m is a positive integer, we may use the relations $z^{-1} = 1/z$ and $1/e^z = e^{-z}$ to write

$$z^{1/n} = (z^{1/m})^{-1} = \left[\exp\left(\frac{1}{m}\log z\right)\right]^{-1}$$

$$= 1\Big/\left[\exp\left(\frac{1}{m}\log z\right)\right] = \exp\left(-\frac{1}{m}\log z\right) = \exp\left(\frac{1}{n}\log z\right).$$

SECTION 32

1. In each part below, $n = 0,\pm1,\pm2,\ldots$.

(a) $(1+i)^i = \exp[i\log(1+i)] = \exp\left\{i\left[\ln\sqrt{2} + i\left(\frac{\pi}{4} + 2n\pi\right)\right]\right\}$

$$= \exp\left[\frac{i}{2}\ln 2 - \left(\frac{\pi}{4} + 2n\pi\right)\right] = \exp\left(-\frac{\pi}{4} - 2n\pi\right)\exp\left(\frac{i}{2}\ln 2\right).$$

Since n takes on all integral values, the term $-2n\pi$ here can be replaced by $+2n\pi$. Thus

$$(1+i)^i = \exp\left(-\frac{\pi}{4} + 2n\pi\right)\exp\left(\frac{i}{2}\ln 2\right).$$

(b) $(-1)^{1/\pi} = \exp\left[\frac{1}{\pi}\log(-1)\right] = \exp\left\{\frac{1}{\pi}[\ln 1 + i(\pi + 2n\pi]\right\} = \exp[(2n+1)i].$

2. *(a)* P.V. $i^i = \exp(i\text{Log}i) = \exp\left[i\left(\ln 1 + i\frac{\pi}{2}\right)\right] = \exp\left(-\frac{\pi}{2}\right).$

(b) P.V. $\left[\frac{e}{2}(-1-\sqrt{3}i)\right]^{3\pi i} = \exp\left\{3\pi i\text{Log}\left[\frac{e}{2}(-1-\sqrt{3}i)\right]\right\} = \exp\left[3\pi i\left(\ln e - i\frac{2\pi}{3}\right)\right]$

$$= \exp(2\pi^2)\exp(i3\pi) = -\exp(2\pi^2).$$

(c) P.V. $(1-i)^{4i} = \exp[4i\text{Log}(1-i)] = \exp\left[4i\left(\ln\sqrt{2} - i\frac{\pi}{4}\right)\right] = e^{\pi}e^{i4\ln\sqrt{2}}$

$$= e^{\pi}[\cos(4\ln\sqrt{2}) + i\sin(4\ln\sqrt{2})] = e^{\pi}[\cos(2\ln 2) + i\sin(2\ln 2)].$$

3. Since $-1+\sqrt{3}i = 2e^{2\pi i/3}$, we may write

$$(-1+\sqrt{3}i)^{3/2} = \exp\left[\frac{3}{2}\log(-1+\sqrt{3}i)\right] = \exp\left\{\frac{3}{2}\left[\ln 2 + i\left(\frac{2\pi}{3} + 2n\pi\right)\right]\right\}$$
$$= \exp[\ln(2^{3/2}) + (3n+1)\pi i] = 2\sqrt{2}\exp[(3n+1)\pi i],$$

where $n = 0, \pm 1, \pm 2, \ldots$. Observe that if n is even, then $3n+1$ is odd; and so $\exp[(3n+1)\pi i] = -1$. On the other hand, if n is odd, $3n+1$ is even; and this means that $\exp[(3n+1)\pi i] = 1$. So only two distinct values of $(-1+\sqrt{3}i)^{3/2}$ arise. Specifically,

$$(-1+\sqrt{3}i)^{3/2} = \pm 2\sqrt{2}.$$

5. We consider here any nonzero complex number z_0 in the exponential form $z_0 = r_0 \exp i\Theta_0$, where $-\pi < \Theta_0 \le \pi$. According to Sec. 8, the principal value of $z^{1/n}$ is $\sqrt[n]{r_0}\exp\left(i\frac{\Theta_0}{n}\right)$; and, according to Sec. 32, that value is

$$\exp\left(\frac{1}{n}\text{Log} z\right) = \exp\left[\frac{1}{n}(\ln r_0 + i\Theta_0)\right] = \exp\left(\ln\sqrt[n]{r_0}\right)\exp\left(i\frac{\Theta_0}{n}\right) = \sqrt[n]{r_0}\exp\left(i\frac{\Theta_0}{n}\right).$$

These two expressions are evidently the same.

7. Observe that when $c = a + bi$ is any fixed complex number, where $c \ne 0, \pm 1, \pm 2, \ldots$, the power i^c can be written as

$$i^c = \exp(c\log i) = \exp\left\{(a+bi)\left[\ln 1 + i\left(\frac{\pi}{2} + 2n\pi\right)\right]\right\}$$
$$= \exp\left[-b\left(\frac{\pi}{2} + 2n\pi\right) + ia\left(\frac{\pi}{2} + 2n\pi\right)\right] \qquad (n = 0, \pm 1, \pm 2, \ldots).$$

Thus

$$|i^c| = \exp\left[-b\left(\frac{\pi}{2} + 2n\pi\right)\right] \qquad (n = 0, \pm 1, \pm 2, \ldots),$$

and it is clear that $|i^c|$ is multiple-valued unless $b = 0$, or c is real. Note that the restriction $c \ne 0, \pm 1, \pm 2, \ldots$ ensures that i^c is multiple-valued even when $b = 0$.

SECTION 33

1. The desired derivatives can be found by writing

$$\frac{d}{dz}\sin z = \frac{d}{dz}\left(\frac{e^{iz}-e^{-iz}}{2i}\right) = \frac{1}{2i}\left(\frac{d}{dz}e^{iz} - \frac{d}{dz}e^{-iz}\right)$$

$$= \frac{1}{2i}\left(ie^{iz} + ie^{-iz}\right) = \frac{e^{iz}+e^{-iz}}{2} = \cos z$$

and

$$\frac{d}{dz}\cos z = \frac{d}{dz}\left(\frac{e^{iz}+e^{-iz}}{2}\right) = \frac{1}{2}\left(\frac{d}{dz}e^{iz} + \frac{d}{dz}e^{-iz}\right)$$

$$= \frac{1}{2}\left(ie^{iz} - ie^{-iz}\right)\cdot\frac{i}{i} = -\frac{e^{iz}-e^{-iz}}{2i} = -\sin z.$$

2. From the expressions

$$\sin z = \frac{e^{iz}-e^{-iz}}{2i} \quad \text{and} \quad \cos z = \frac{e^{iz}+e^{-iz}}{2},$$

we see that

$$\cos z + i\sin z = \frac{e^{iz}+e^{-iz}}{2} + \frac{e^{iz}-e^{-iz}}{2} = e^{iz}.$$

3. Equation (4), Sec. 33 is

$$2\sin z_1 \cos z_2 = \sin(z_1+z_2) + \sin(z_1-z_2).$$

Interchanging z_1 and z_2 here and using the fact that sin z is an odd function, we have

$$2\cos z_1 \sin z_2 = \sin(z_1+z_2) - \sin(z_1-z_2).$$

Addition of corresponding sides of these two equations now yields

$$2(\sin z_1 \cos z_2 + \cos z_1 \sin z_2) = 2\sin(z_1+z_2),$$

or

$$\sin(z_1+z_2) = \sin z_1 \cos z_2 + \cos z_1 \sin z_2.$$

4. Differentiating each side of equation (5), Sec. 33, with respect to z_1, we have

$$\cos(z_1+z_2) = \cos z_1 \cos z_2 - \sin z_1 \sin z_2.$$

7. *(a)* From the identity $\sin^2 z + \cos^2 z = 1$, we have

$$\frac{\sin^2 z}{\cos^2 z} + \frac{\cos^2 z}{\cos^2 z} = \frac{1}{\cos^2 z}, \quad \text{or} \quad 1 + \tan^2 z = \sec^2 z.$$

(b) Also,

$$\frac{\sin^2 z}{\sin^2 z} + \frac{\cos^2 z}{\sin^2 z} = \frac{1}{\sin^2 z}, \quad \text{or} \quad 1 + \cot^2 z = \csc^2 z.$$

9. From the expression

$$\sin z = \sin x \cosh y + i \cos x \sinh y,$$

we find that

$$\begin{aligned} |\sin z|^2 &= \sin^2 x \cosh^2 y + \cos^2 x \sinh^2 y \\ &= \sin^2 x(1 + \sinh^2 y) + (1 - \sin^2 x)\sinh^2 y \\ &= \sin^2 x + \sinh^2 y. \end{aligned}$$

The expression

$$\cos z = \cos x \cosh y + i \sin x \sinh y,$$

on the other hand, tells us that

$$\begin{aligned} |\cos z|^2 &= \cos^2 x \cosh^2 y + \sin^2 x \sinh^2 y \\ &= \cos^2 x(1 + \sinh^2 y) + (1 - \cos^2 x)\sinh^2 y \\ &= \cos^2 x + \sinh^2 y. \end{aligned}$$

10. Since $\sinh^2 y$ is never negative, it follows from Exercise 9 that

(a) $$|\sin z|^2 \geq \sin^2 x, \quad \text{or} \quad |\sin z| \geq |\sin x|$$

and that

(b) $$|\cos z|^2 \geq \cos^2 x, \quad \text{or} \quad |\cos z| \geq |\cos x|.$$

11. In this problem we shall use the identities

$$|\sin z|^2 = \sin^2 x + \sinh^2 y, \quad |\cos z|^2 = \cos^2 x + \sinh^2 y.$$

(a) Observe that

$$\sinh^2 y = |\sin z|^2 - \sin^2 x \le |\sin z|^2$$

and

$$|\sin z|^2 = \sin^2 x + (\cosh^2 y - 1) = \cosh^2 y - (1 - \sin^2 x)$$
$$= \cosh^2 y - \cos^2 x \le \cosh^2 y.$$

Thus

$$\sinh^2 y \le |\sin z|^2 \le \cosh^2 y, \quad \text{or} \quad |\sinh y| \le |\sin z| \le \cosh y.$$

(b) On the other hand,

$$\sinh^2 y = |\cos z|^2 - \cos^2 x \le |\cos z|^2$$

and

$$|\cos z|^2 = \cos^2 x + (\cosh^2 y - 1) = \cosh^2 y - (1 - \cos^2 x)$$
$$= \cosh^2 y - \sin^2 x \le \cosh^2 y.$$

Hence

$$\sinh^2 y \le |\cos z|^2 \le \cosh^2 y, \quad \text{or} \quad |\sinh y| \le |\cos z| \le \cosh y.$$

13. By writing $f(z) = \sin \bar{z} = \sin(x - iy) = \sin x \cosh y - i \cos x \sinh y$, we have

$$f(z) = u(x,y) + iv(x,y),$$

where

$$u(x,y) = \sin x \cosh y \quad \text{and} \quad v(x,y) = -\cos x \sinh y.$$

If the Cauchy-Riemann equations $u_x = v_y$, $u_y = -v_x$ are to hold, it is easy to see that

$$\cos x \cosh y = 0 \quad \text{and} \quad \sin x \sinh y = 0.$$

Since $\cosh y$ is never zero, it follows from the first of these equations that $\cos x = 0$; that is, $x = \dfrac{\pi}{2} + n\pi$ $(n = 0 \pm 1, \pm 2, \ldots)$. Furthermore, since $\sin x$ is nonzero for each of these values of x, the second equation tells us that $\sinh y = 0$, or $y = 0$. Thus the Cauchy-Riemann equations hold only at the points

$$z = \frac{\pi}{2} + n\pi \qquad (n = 0 \pm 1, \pm 2, \ldots).$$

Evidently, then, there is no neighborhood of any point throughout which f is analytic, and we may conclude that $\sin \bar{z}$ is not analytic anywhere.

The function $f(z) = \cos \bar{z} = \cos(x - iy) = \cos x \cosh y + i \sin x \sinh y$ can be written as

$$f(z) = u(x,y) + iv(x,y),$$

where

$$u(x,y) = \cos x \cosh y \quad \text{and} \quad v(x,y) = \sin x \sinh y.$$

If the Cauchy-Riemann equations $u_x = v_y$, $u_y = -v_x$ hold, then

$$\sin x \cosh y = 0 \quad \text{and} \quad \cos x \sinh y = 0.$$

The first of these equations tells us that $\sin x = 0$, or $x = n\pi \ (n = 0, \pm 1, \pm 2, \ldots)$. Since $\cos n\pi \neq 0$, it follows that $\sinh y = 0$, or $y = 0$. Consequently, the Cauchy-Riemann equations hold only when

$$z = n\pi \qquad (n = 0 \pm 1, \pm 2, \ldots).$$

So there is no neighborhood throughout which f is analytic, and this means that $\cos \bar{z}$ is nowhere analytic.

16. *(a)* Use expression (12), Sec. 33, to write

$$\overline{\cos(iz)} = \overline{\cos(-y + ix)} = \cos y \cosh x - i \sin y \sinh x$$

and

$$\cos(i\bar{z}) = \cos(y + ix) = \cos y \cosh x - i \sin y \sinh x.$$

This shows that $\overline{\cos(iz)} = \cos(i\bar{z})$ for all z.

(b) Use expression (11), Sec. 33, to write

$$\overline{\sin(iz)} = \overline{\sin(-y + ix)} = -\sin y \cosh x - i \cos y \sinh x$$

and

$$\sin(i\bar{z}) = \sin(y + ix) = \sin y \cosh x + i \cos y \sinh x.$$

Evidently, then, the equation $\overline{\sin(iz)} = \sin(i\bar{z})$ is equivalent to the pair of equations

$$\sin y \cosh x = 0, \quad \cos y \sinh x = 0.$$

Since $\cosh x$ is never zero, the first of these equations tells us that $\sin y = 0$. Consequently, $y = n\pi \quad (n = 0, \pm 1, \pm 2, \ldots)$. Since $\cos n\pi = (-1)^n \neq 0$, the second equation tells us that $\sinh x = 0$, or that $x = 0$. So we may conclude that $\overline{\sin(iz)} = \sin(i\bar{z})$ if and only if $z = 0 + in\pi = n\pi i \ (n = 0, \pm 1, \pm 2, \ldots)$.

17. Rewriting the equation $\sin z = \cosh 4$ as $\sin x \cosh y + i \cos x \sinh y = \cosh 4$, we see that we need to solve the pair of equations

$$\sin x \cosh y = \cosh 4, \quad \cos x \sinh y = 0$$

for x and y. If $y=0$, the first equation becomes $\sin x = \cosh 4$, which cannot be satisfied by any x since $\sin x \le 1$ and $\cosh 4 > 1$. So $y \ne 0$, and the second equation requires that $\cos x = 0$. Thus

$$x = \frac{\pi}{2} + n\pi \qquad (n = 0 \pm 1, \pm 2, \ldots).$$

Since

$$\sin\left(\frac{\pi}{2} + n\pi\right) = (-1)^n,$$

the first equation then becomes $(-1)^n \cosh y = \cosh 4$, which cannot hold when n is odd. If n is even, it follows that $y = \pm 4$. Finally, then, the roots of $\sin z = \cosh 4$ are

$$z = \left(\frac{\pi}{2} + 2n\pi\right) \pm 4i \qquad (n = 0 \pm 1, \pm 2, \ldots).$$

18. The problem here is to find all roots of the equation $\cos z = 2$. We start by writing that equation as $\cos x \cosh y - i \sin x \sinh y = 2$. Thus we need to solve the pair of equations

$$\cos x \cosh y = 2, \quad \sin x \sinh y = 0$$

for x and y. We note that $y \ne 0$ since $\cos x = 2$ if $y = 0$, and that is impossible. So the second in the pair of equations to be solved tells us that $\sin x = 0$, or that $x = n\pi$ $(n = 0 \pm 1, \pm 2, \ldots)$. The first equation then tells us that $(-1)^n \cosh y = 2$; and, since $\cosh y$ is always positive, n must be even. That is, $x = 2n\pi$ $(n = 0 \pm 1, \pm 2, \ldots)$. But this means that $\cosh y = 2$, or $y = \cosh^{-1} 2$. Consequently, the roots of the given equation are

$$z = 2n\pi + i\cosh^{-1} 2 \qquad (n = 0 \pm 1, \pm 2, \ldots).$$

To express $\cosh^{-1} 2$, which has two values, in a different way, we begin with $y = \cosh^{-1} 2$, or $\cosh y = 2$. This tells us that $e^y + e^{-y} = 4$; and, rewriting this as

$$(e^y)^2 - 4(e^y) + 1 = 0,$$

we may apply the quadratic formula to obtain $e^y = 2 \pm \sqrt{3}$, or $y = \ln(2 \pm \sqrt{3})$. Finally, with the observation that

$$\ln(2-\sqrt{3}) = \ln\left[\frac{(2-\sqrt{3})(2+\sqrt{3})}{2+\sqrt{3}}\right] = \ln\left(\frac{1}{2+\sqrt{3}}\right) = -\ln(2+\sqrt{3}),$$

we arrive at this alternative form of the roots:

$$z = 2n\pi \pm i\ln(2+\sqrt{3}) \qquad (n = 0 \pm 1, \pm 2, \ldots).$$

SECTION 34

1. To find the derivatives of sinh$z$ and cosh$z$, we write

$$\frac{d}{dz}\sinh z = \frac{d}{dz}\left(\frac{e^z - e^{-z}}{2}\right) = \frac{1}{2}\frac{d}{dz}(e^z - e^{-z}) = \frac{e^z + e^{-z}}{2} = \cosh z$$

and

$$\frac{d}{dz}\cosh z = \frac{d}{dz}\left(\frac{e^z + e^{-z}}{2}\right) = \frac{1}{2}\frac{d}{dz}(e^z + e^{-z}) = \frac{e^z - e^{-z}}{2} = \sinh z.$$

3. Identity (7), Sec. 33, is $\sin^2 z + \cos^2 z = 1$. Replacing z by iz here and using the identities

$$\sin(iz) = i\sinh z \quad \text{and} \quad \cos(iz) = \cosh z,$$

we find that $i^2 \sinh^2 z + \cosh^2 z = 1$, or

$$\cosh^2 z - \sinh^2 z = 1.$$

Identity (6), Sec. 33, is $\cos(z_1 + z_2) = \cos z_1 \cos z_2 - \sin z_1 \sin z_2$. Replacing z_1 by iz_1 and z_2 by iz_2 here, we have $\cos[i(z_1 + z_2)] = \cos(iz_1)\cos(iz_2) - \sin(iz_1)\sin(iz_2)$. The same identities that were used just above then lead to

$$\cosh(z_1 + z_2) = \cosh z_1 \cosh z_2 + \sinh z_1 \sinh z_2.$$

6. We wish to show that

$$|\sinh x| \le |\cosh z| \le \cosh x$$

in two different ways.

(a) Identity (12), Sec. 34, is $|\cosh z|^2 = \sinh^2 x + \cos^2 y$. Thus $|\cosh z|^2 - \sinh^2 x \ge 0$; and this tells us that $\sinh^2 x \le |\cosh z|^2$, or $|\sinh x| \le |\cosh z|$. On the other hand, since $|\cosh z|^2 = (\cosh^2 x - 1) + \cos^2 y = \cosh^2 x - (1 - \cos^2 y) = \cosh^2 x - \sin^2 y$, we know that $|\cosh z|^2 - \cosh^2 x \le 0$. Consequently, $|\cosh z|^2 \le \cosh^2 x$, or $|\cosh z| \le \cosh x$.

(b) Exercise 11(*b*), Sec. 33, tells us that $|\sinh y| \le |\cos z| \le \cosh y$. Replacing z by iz here and recalling that $\cos iz = \cosh z$ and $iz = -y + ix$, we obtain the desired inequalities.

7. *(a)* Observe that

$$\sinh(z + \pi i) = \frac{e^{z+\pi i} - e^{-(z+\pi i)}}{2} = \frac{e^z e^{\pi i} - e^{-z} e^{-\pi i}}{2} = \frac{-e^z + e^{-z}}{2} = -\frac{e^z - e^{-z}}{2} = -\sinh z.$$

(b) Also,

$$\cosh(z+\pi i)=\frac{e^{z+\pi i}+e^{-(z+\pi i)}}{2}=\frac{e^{z}e^{\pi i}+e^{-z}e^{-\pi i}}{2}=\frac{-e^{z}-e^{-z}}{2}=-\frac{e^{z}+e^{-z}}{2}=-\cosh z.$$

(c) From parts *(a)* and *(b)*, we find that

$$\tanh(z+\pi i)=\frac{\sinh(z+\pi i)}{\cosh(z+\pi i)}=\frac{-\sinh z}{-\cosh z}=\frac{\sinh z}{\cosh z}=\tanh z.$$

9. The zeros of the hyperbolic tangent function

$$\tanh z=\frac{\sinh z}{\cosh z}$$

are the same as the zeros of $\sinh z$, which are $z=n\pi i\ (n=0,\pm1,\pm2,\ldots)$. The singularities of $\tanh z$ are the zeros of $\cosh z$, or $z=\left(\frac{\pi}{2}+n\pi\right)i\ (n=0,\pm1,\pm2,\ldots)$.

15. *(a)* Observe that, since $\sinh z=i$ can be written as $\sinh x\cos y+i\cosh x\sin y=i$, we need to solve the pair of equations

$$\sinh x\cos y=0,\quad \cosh x\sin y=1.$$

If $x=0$, the second of these equations becomes $\sin y=1$; and so $y=\frac{\pi}{2}+2n\pi$ $(n=0,\pm1,\pm2,\ldots)$. Hence

$$z=\left(2n+\frac{1}{2}\right)\pi i \qquad (n=0,\pm1,\pm2,\ldots).$$

If $x\neq0$, the first equation requires that $\cos y=0$, or $y=\frac{\pi}{2}+n\pi$ $(n=0,\pm1,\pm2,\ldots)$. The second then becomes $(-1)^n\cosh x=1$. But there is no nonzero value of x satisfying this equation, and we have no additional roots of $\sinh z=i$.

(b) Rewriting $\cosh z=\frac{1}{2}$ as $\cosh x\cos y+i\sinh x\sin y=\frac{1}{2}$, we see that x and y must satisfy the pair of equations

$$\cosh x\cos y=\frac{1}{2},\quad \sinh x\sin y=0.$$

If $x=0$, the second equation is satisfied and the first equation becomes $\cos y = \frac{1}{2}$. Thus $y = \cos^{-1}\frac{1}{2} = \pm\frac{\pi}{3} + 2n\pi$ $(n = 0, \pm1, \pm2, \ldots)$, and this means that

$$z = \left(2n \pm \frac{1}{3}\right)\pi i \qquad (n = 0, \pm1, \pm2, \ldots).$$

If $x \neq 0$, the second equation tells us that $y = n\pi$ $(n = 0, \pm1, \pm2, \ldots)$. The first then becomes $(-1)^n \cosh x = \frac{1}{2}$. But this equation in x has no solution since $\cosh x \geq 1$ for all x. Thus no additional roots of $\cosh z = \frac{1}{2}$ are obtained.

16. Let us rewrite $\cosh z = -2$ as $\cosh x \cos y + i \sinh x \sin y = -2$. The problem is evidently to solve the pair of equations

$$\cosh x \cos y = -2, \quad \sinh x \sin y = 0.$$

If $x=0$, the second equation is satisfied and the first reduces to $\cos y = -2$. Since there is no y satisfying this equation, no roots of $\cosh z = -2$ arise.
If $x \neq 0$, we find from the second equation that $\sin y = 0$, or $y = n\pi$ $(n = 0, \pm1, \pm2, \ldots)$. Since $\cos n\pi = (-1)^n$, it follows from the first equation that $(-1)^n \cosh x = -2$. But this equation can hold only when n is odd, in which case $x = \cosh^{-1} 2$. Consequently,

$$z = \cosh^{-1} 2 + (2n+1)\pi i \qquad (n = 0, \pm1, \pm2, \ldots).$$

Recalling from the solution of Exercise 18, Sec 33, that $\cosh^{-1} 2 = \pm\ln(2+\sqrt{3})$, we note that these roots can also be written as

$$z = \pm\ln(2+\sqrt{3}) + (2n+1)\pi i \qquad (n = 0, \pm1, \pm2, \ldots).$$

Chapter 4

SECTION 37

2. *(a)* $\displaystyle\int_1^2\left(\frac{1}{t}-i\right)^2 dt=\int_1^2\left(\frac{1}{t^2}-1\right)dt-2i\int_1^2\frac{dt}{t}=-\frac{1}{2}-2i\ln 2=-\frac{1}{2}-i\ln 4;$

(b) $\displaystyle\int_0^{\pi/6} e^{i2t}dt=\left[\frac{e^{i2t}}{2i}\right]_0^{\pi/6}=\frac{1}{2i}\left[\cos\frac{\pi}{3}+i\sin\frac{\pi}{3}-1\right]=\frac{\sqrt{3}}{4}+\frac{i}{4};$

(c) Since $|e^{-bz}|=e^{-bx}$, we find that

$$\int_0^\infty e^{-zt}dt=\lim_{b\to\infty}\int_0^b e^{-zt}dt=\lim_{b\to\infty}\left[\frac{e^{-zt}}{-z}\right]_{t=0}^{t=b}=\frac{1}{z}\lim_{b\to\infty}\left(1-e^{-bz}\right)=\frac{1}{z}\quad\text{when } \operatorname{Re} z>0.$$

3. The problem here is to verify that

$$\int_0^{2\pi} e^{im\theta}e^{-in\theta}d\theta=\begin{cases}0 & \text{when } m\neq n,\\ 2\pi & \text{when } m=n.\end{cases}$$

To do this, we write

$$I=\int_0^{2\pi} e^{im\theta}e^{-in\theta}d\theta=\int_0^{2\pi} e^{i(m-n)\theta}d\theta$$

and observe that when $m\neq n$,

$$I=\left[\frac{e^{i(m-n)\theta}}{i(m-n)}\right]_0^{2\pi}=\frac{1}{i(m-n)}-\frac{1}{i(m-n)}=0.$$

When $m=n$, I becomes

$$I=\int_0^{2\pi} d\theta=2\pi;$$

and the verification is complete.

4. First of all,

$$\int_0^\pi e^{(1+i)x}dx=\int_0^\pi e^x\cos x\,dx+i\int_0^\pi e^x\sin x\,dx.$$

But also,

$$\int_0^\pi e^{(1+i)x}dx=\left[\frac{e^{(1+i)x}}{1+i}\right]_0^\pi=\frac{e^\pi e^{i\pi}-1}{1+i}=\frac{-e^\pi-1}{1+i}\cdot\frac{1-i}{1-i}=-\frac{1+e^\pi}{2}+i\frac{1+e^\pi}{2}.$$

Equating the real parts and then the imaginary parts of these two expressions, we find that

$$\int_0^\pi e^x \cos x \, dx = -\frac{1+e^\pi}{2} \quad \text{and} \quad \int_0^\pi e^x \sin x \, dx = \frac{1+e^\pi}{2}.$$

5. Consider the function $w(t) = e^{it}$ and observe that

$$\int_0^{2\pi} w(t)dt = \int_0^{2\pi} e^{it} dt = \left[\frac{e^{it}}{i}\right]_0^{2\pi} = \frac{1}{i} - \frac{1}{i} = 0.$$

Since $|w(c)(2\pi - 0)| = |e^{ic}| 2\pi = 2\pi$ for every real number c, it is clear that there is no number c in the interval $0 < t < 2\pi$ such that

$$\int_0^{2\pi} w(t)dt = w(c)(2\pi - 0).$$

6. *(a)* Suppose that $w(t)$ is even. It is straightforward to show that $u(t)$ and $v(t)$ must be even. Thus

$$\int_{-a}^{a} w(t)dt = \int_{-a}^{a} u(t)dt + i\int_{-a}^{a} v(t)dt = 2\int_0^a u(t)dt + 2i\int_0^a v(t)dt$$

$$= 2\left[\int_0^a u(t)dt + i\int_0^a v(t)dt\right] = 2\int_0^a w(t)dt.$$

(b) Suppose, on the other hand, that $w(t)$ is odd. It follows that $u(t)$ and $v(t)$ are odd, and so

$$\int_{-a}^{a} w(t)dt = \int_{-a}^{a} u(t)dt + i\int_{-a}^{a} v(t)dt = 0 + i0 = 0.$$

7. Consider the functions

$$P_n(x) = \frac{1}{\pi}\int_0^\pi \left(x + i\sqrt{1-x^2}\cos\theta\right)^n d\theta \qquad (n = 0,1,2,\ldots),$$

where $-1 \le x \le 1$. Since

$$\left|x + i\sqrt{1-x^2}\cos\theta\right| = \sqrt{x^2 + (1-x^2)\cos^2\theta} \le \sqrt{x^2 + (1-x^2)} = 1,$$

it follows that

$$|P_n(x)| \le \frac{1}{\pi}\int_0^\pi \left|x + i\sqrt{1-x^2}\cos\theta\right|^n d\theta \le \frac{1}{\pi}\int_0^\pi d\theta = 1.$$

SECTION 38

1. *(a)* Start by writing

$$I = \int_{-b}^{-a} w(-t)dt = \int_{-b}^{-a} u(-t)dt + i\int_{-b}^{-a} v(-t)dt.$$

The substitution $\tau = -t$ in each of these two integrals on the right then yields

$$I = -\int_{b}^{a} u(\tau)d\tau - i\int_{b}^{a} v(\tau)d\tau = \int_{a}^{b} u(\tau)d\tau + i\int_{a}^{b} v(\tau)d\tau = \int_{a}^{b} w(\tau)d\tau.$$

That is,

$$\int_{-b}^{-a} w(-t)dt = \int_{a}^{b} w(\tau)d\tau.$$

(b) Start with

$$I = \int_{a}^{b} w(t)dt = \int_{a}^{b} u(t)dt + i\int_{a}^{b} v(t)dt$$

and then make the substitution $t = \varphi(\tau)$ in each of the integrals on the right. The result is

$$I = \int_{\alpha}^{\beta} u[\phi(\tau)]\phi'(\tau)d\tau + i\int_{\alpha}^{\beta} v[\phi(\tau)]\phi'(\tau)d\tau = \int_{\alpha}^{\beta} w[\phi(\tau)]\phi'(\tau)d\tau.$$

That is,

$$\int_{a}^{b} w(t)dt = \int_{\alpha}^{\beta} w[\phi(\tau)]\phi'(\tau)d\tau.$$

3. The slope of the line through the points (α, a) and (β, b) in the τt plane is

$$m = \frac{b-a}{\beta-\alpha}.$$

So the equation of that line is

$$t - a = \frac{b-a}{\beta-\alpha}(\tau-\alpha).$$

Solving this equation for t, one can rewrite it as

$$t = \frac{b-a}{\beta-\alpha}\tau + \frac{a\beta - b\alpha}{\beta-\alpha}.$$

Since $t = \phi(\tau)$, then,

$$\phi(\tau) = \frac{b-a}{\beta-\alpha}\tau + \frac{a\beta - b\alpha}{\beta-\alpha}.$$

4. If $Z(\tau) = z[\phi(\tau)]$, where $z(t) = x(t) + iy(t)$ and $t = \phi(\tau)$, then

$$Z(\tau) = x[\phi(\tau)] + iy[\phi(\tau)].$$

Hence

$$Z'(\tau) = \frac{d}{d\tau}x[\phi(\tau)] + i\frac{d}{d\tau}y[\phi(\tau)] = x'[\phi(\tau)]\phi'(\tau) + iy'[\phi(\tau)]\phi'(\tau)$$

$$= \{x'[\phi(\tau)] + iy'[\phi(\tau)]\}\phi'(\tau) = z'[\phi(\tau)]\phi'(\tau).$$

5. If $w(t) = f[z(t)]$ and $f(z) = u(x,y) + iv(x,y)$, $z(t) = x(t) + iy(t)$, we have

$$w(t) = u[x(t), y(t)] + iv[x(t), y(t)].$$

The chain rule tells us that

$$\frac{du}{dt} = u_x x' + u_y y' \quad \text{and} \quad \frac{dv}{dt} = v_x x' + v_y y',$$

and so

$$w'(t) = (u_x x' + u_y y') + i(v_x x' + v_y y').$$

In view of the Cauchy-Riemann equations $u_x = v_y$ and $u_y = -v_x$, then,

$$w'(t) = (u_x x' - v_x y') + i(v_x x' + u_x y') = (u_x + iv_x)(x' + iy').$$

That is,

$$w'(t) = \{u_x[x(t), y(t)] + iv_x[x(t), y(t)]\}[x'(t) + iy'(t)] = f'[z(t)]z'(t)$$

when $t = t_0$.

SECTION 40

1. (*a*) Let C be the semicircle $z = 2e^{i\theta}$ $(0 \le \theta \le \pi)$, shown below.

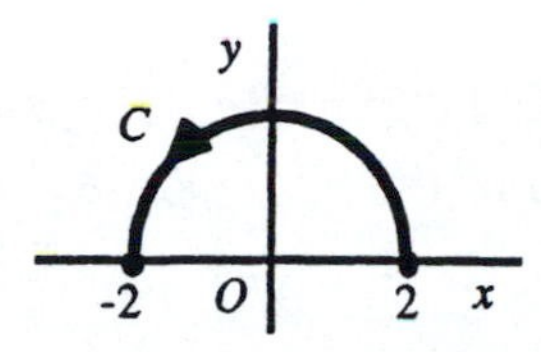

Then

$$\int_C \frac{z+2}{z}\,dz = \int_C \left(1+\frac{2}{z}\right)dz = \int_0^{\pi}\left(1+\frac{2}{2e^{i\theta}}\right)2ie^{i\theta}\,d\theta = 2i\int_0^{\pi}(e^{i\theta}+1)\,d\theta$$

$$= 2i\left[\frac{e^{i\theta}}{i}+\theta\right]_0^{\pi} = 2i(i+\pi+i) = -4+2\pi i.$$

(*b*) Now let C be the semicircle $z = 2e^{i\theta}$ $(\pi \le \theta \le 2\pi)$ just below.

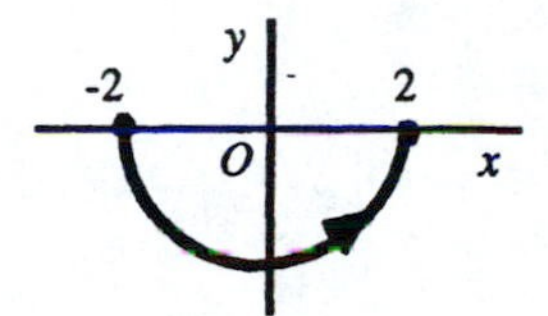

This is the same as part (*a*), except for the limits of integration. Thus

$$\int_C \frac{z+2}{z}\,dz = 2i\left[\frac{e^{i\theta}}{i}+\theta\right]_{\pi}^{2\pi} = 2i(-i+2\pi-i-\pi) = 4+2\pi i.$$

(*c*) Finally, let C denote the entire circle $z = 2e^{i\theta}$ $(0 \le \theta \le 2\pi)$. In this case,

$$\int_C \frac{z+2}{z}\,dz = 4\pi i,$$

the value here being the sum of the values of the integrals in parts (*a*) and (*b*).

2. (*a*) The arc is $C: z = 1+e^{i\theta}$ $(\pi \le \theta \le 2\pi)$. Then

$$\int_C (z-1)\,dz = \int_{\pi}^{2\pi}(1+e^{i\theta}-1)ie^{i\theta}\,d\theta = i\int_{\pi}^{2\pi}e^{i2\theta}\,d\theta = i\left[\frac{e^{i2\theta}}{2i}\right]_{\pi}^{2\pi}$$

$$= \frac{1}{2}\left(e^{i4\pi}-e^{i2\pi}\right) = \frac{1}{2}(1-1) = 0.$$

(b) Here $C: z = x\ (0 \le x \le 2)$. Then

$$\int_C (z-1)\,dz = \int_0^2 (x-1)\,dx = \left[\frac{x^2}{2} - x\right]_0^2 = 0.$$

3. In this problem, the path C is the sum of the paths C_1, C_2, C_3, and C_4 that are shown below.

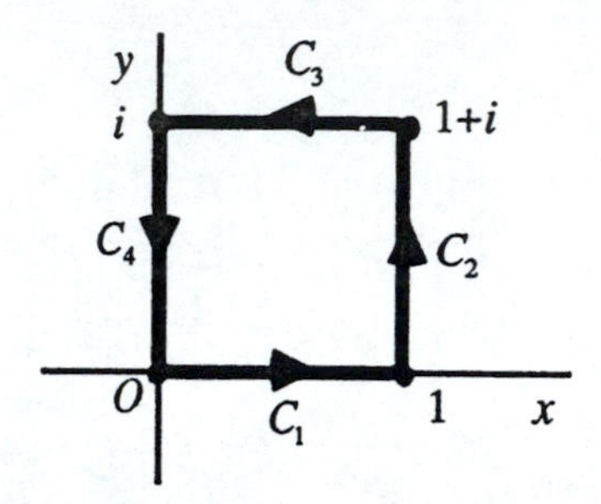

The function to be integrated around the closed path C is $f(z) = \pi e^{\pi \bar{z}}$. We observe that $C = C_1 + C_2 + C_3 + C_4$ and find the values of the integrals along the individual legs of the square C.

(i) Since C_1 is $z = x\ (0 \le x \le 1)$,

$$\int_{C_1} \pi e^{\pi \bar{z}} dz = \pi \int_0^1 e^{\pi x} dx = e^{\pi} - 1.$$

(ii) Since C_2 is $z = 1 + iy\ (0 \le y \le 1)$,

$$\int_{C_2} \pi e^{\pi \bar{z}} dz = \pi \int_0^1 e^{\pi(1-iy)} i\,dy = e^{\pi} \pi i \int_0^1 e^{-i\pi y} dy = 2e^{\pi}.$$

(iii) Since C_3 is $z = (1-x) + i\ (0 \le x \le 1)$,

$$\int_{C_3} \pi e^{\pi \bar{z}} dz = \pi \int_0^1 e^{\pi[(1-x)-i]}(-1)dx = \pi e^{\pi} \int_0^1 e^{-\pi x} dx = e^{\pi} - 1.$$

(iv) Since C_4 is $z = i(1-y)\ (0 \le y \le 1)$,

$$\int_{C_4} \pi e^{\pi \bar{z}} dz = \pi \int_0^1 e^{-\pi(1-y)i}(-i)dy = \pi i \int_0^1 e^{i\pi y} dy = -2.$$

Finally, then, since

$$\int_C \pi e^{\pi \bar{z}} dz = \int_{C_1} \pi e^{\pi \bar{z}} dz + \int_{C_2} \pi e^{\pi \bar{z}} dz + \int_{C_3} \pi e^{\pi \bar{z}} dz + \int_{C_4} \pi e^{\pi \bar{z}} dz,$$

we find that

$$\int_C \pi e^{\pi \bar{z}} dz = 4(e^{\pi} - 1).$$

4. The path C is the sum of the paths

$$C_1 : z = x + ix^3 \ (-1 \le x \le 0) \quad \text{and} \quad C_2 : z = x + ix^3 \ (0 \le x \le 1).$$

Using

$$f(z) = 1 \text{ on } C_1 \quad \text{and} \quad f(z) = 4y = 4x^3 \text{ on } C_2,$$

we have

$$\int_C f(z)dz = \int_{C_1} f(z)dz + \int_{C_2} f(z)dz = \int_{-1}^{0} 1(1 + i3x^2)dx + \int_0^1 4x^3(1 + i3x^2)dx$$

$$= \int_{-1}^{0} dx + 3i\int_{-1}^{0} x^2 dx + 4\int_0^1 x^3 dx + 12i\int_0^1 x^5 dx$$

$$= [x]_{-1}^{0} + i[x^3]_{-1}^{0} + [x^4]_0^1 + 2i[x^6]_0^1 = 1 + i + 1 + 2i = 2 + 3i.$$

5. The contour C has some parametric representation $z = z(t)\ (a \le t \le b)$, where $z(a) = z_1$ and $z(b) = z_2$. Then

$$\int_C dz = \int_a^b z'(t)dt = [z(t)]_a^b = z(b) - z(a) = z_2 - z_1.$$

6. To integrate the branch

$$z^{-1+i} = e^{(-1+i)\log z} \qquad (|z| > 0,\ 0 < \arg z < 2\pi)$$

around the circle $C: z = e^{i\theta}\ (0 \le \theta \le 2\pi)$, write

$$\int_C z^{-1+i}\,dz = \int_C e^{(-1+i)\log z}\,dz = \int_0^{2\pi} e^{(-1+i)(\ln 1 + i\theta)}\, ie^{i\theta}d\theta = i\int_0^{2\pi} e^{-i\theta-\theta}\, e^{i\theta}d\theta = i\int_0^{2\pi} e^{-\theta}d\theta = i\left(1 - e^{-2\pi}\right).$$

7. Let C be the positively oriented circle $|z| = 1$, with parametric representation $z = e^{i\theta}\ (0 \le \theta \le 2\pi)$, and let m and n be integers. Then

$$\int_C z^m \bar{z}^n dz = \int_0^{2\pi} \left(e^{i\theta}\right)^m \left(e^{-i\theta}\right)^n ie^{i\theta}d\theta = i\int_0^{2\pi} e^{i(m+1)\theta} e^{-in\theta}d\theta.$$

But we know from Exercise 3, Sec. 37, that

$$\int_0^{2\pi} e^{im\theta} e^{-in\theta} d\theta = \begin{cases} 0 & \text{when} \quad m \ne n, \\ 2\pi & \text{when} \quad m = n. \end{cases}$$

Consequently,

$$\int_C z^m \bar{z}^n dz = \begin{cases} 0 & \text{when} \quad m+1 \neq n, \\ 2\pi i & \text{when} \quad m+1 = n. \end{cases}$$

8. Note that C is the right-hand half of the circle $x^2+y^2=4$. So, on C, $x=\sqrt{4-y^2}$. This suggests the parametric representation $C: z=\sqrt{4-y^2}+iy$ $(-2 \leq y \leq 2)$, to be used here. With that representation, we have

$$\int_C \bar{z}\, dz = \int_{-2}^{2}\left(\sqrt{4-y^2}-iy\right)\left(\frac{-y}{\sqrt{4-y^2}}+i\right)dy$$

$$= \int_{-2}^{2}(-y+y)\,dy + i\int_{-2}^{2}\left(\frac{y^2}{\sqrt{4-y^2}}+\sqrt{4-y^2}\right)dy$$

$$= i\int_{-2}^{2}\frac{y^2+4-y^2}{\sqrt{4-y^2}}\,dy = 4i\int_{-2}^{2}\frac{dy}{\sqrt{4-y^2}} = 4i\left[\sin^{-1}\left(\frac{y}{2}\right)\right]_{-2}^{2}$$

$$= 4i\left[\sin^{-1}(1)-\sin^{-1}(-1)\right] = 4i\left[\frac{\pi}{2}-\left(-\frac{\pi}{2}\right)\right] = 4\pi i.$$

10. Let C_0 be the circle $z=z_0+Re^{i\theta}$ $(-\pi \leq \theta \leq \pi)$.

(a) $$\int_{C_0}\frac{dz}{z-z_0} = \int_{-\pi}^{\pi}\frac{1}{Re^{i\theta}}Rie^{i\theta}\,d\theta = i\int_{-\pi}^{\pi}d\theta = 2\pi i.$$

(b) When $n=\pm1,\pm2,\ldots,$

$$\int_{C_0}(z-z_0)^{n-1}dz = \int_{-\pi}^{\pi}\left(Re^{i\theta}\right)^{n-1}Rie^{i\theta}\,d\theta = iR^n\int_{-\pi}^{\pi}e^{in\theta}\,d\theta$$

$$= \frac{R^n}{n}\left(e^{in\pi}-e^{-in\pi}\right) = i\frac{2R^n}{n}\sin n\pi = 0.$$

11. In this case, where a is any real number other than zero, the same steps as in Exercise 10*(b)*, with a instead of n, yield the result

$$\int_{C_0}(z-z_0)^{a-1}dz = i\frac{2R^a}{a}\sin(a\pi).$$

12. (*a*) The function $f(z)$ is continuous on a smooth arc C, which has a parametric representation $z = z(t)$ $(a \le t \le b)$. Exercise 1(*b*), Sec. 38, enables us to write

$$\int_a^b f[z(t)]z'(t)dt = \int_\alpha^\beta f[Z(\tau)]z'[\phi(\tau)]\phi'(\tau)d\tau,$$

where

$$Z(\tau) = z[\phi(\tau)] \qquad (\alpha \le \tau \le \beta).$$

But expression (14), Sec 38, tells us that

$$z'[\phi(\tau)]\phi'(\tau) = Z'(\tau);$$

and so

$$\int_a^b f[z(t)]z'(t)dt = \int_\alpha^\beta f[Z(\tau)]Z'(\tau)d\tau.$$

(*b*) Suppose that C is any contour and that $f(z)$ is piecewise continuous on C. Since C can be broken up into a finite chain of smooth arcs on which $f(z)$ is continuous, the identity obtained in part (*a*) remains valid.

SECTION 41

1. Let C be the arc of the circle $|z| = 2$ shown below.

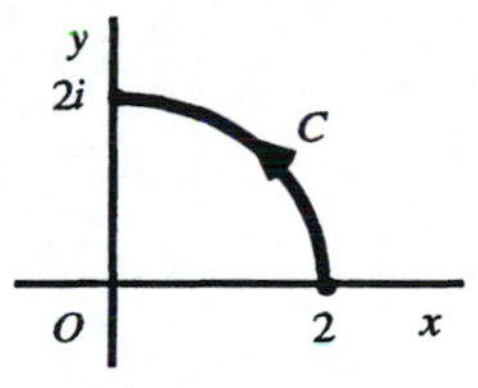

Without evaluating the integral, let us find an upper bound for $\left|\int_C \frac{dz}{z^2-1}\right|$. To do this, we note that if z is a point on C,

$$|z^2 - 1| \ge ||z^2| - 1| = ||z|^2 - 1| = |4 - 1| = 3.$$

Thus

$$\left|\frac{1}{z^2-1}\right| = \frac{1}{|z^2-1|} \le \frac{1}{3}.$$

Also, the length of C is $\frac{1}{4}(4\pi) = \pi$. So, taking $M = \frac{1}{3}$ and $L = \pi$, we find that

$$\left|\int_C \frac{dz}{z^2 - 1}\right| \le ML = \frac{\pi}{3}.$$

2. The path C is as shown in the figure below. The midpoint of C is clearly the closest point on C to the origin. The distance of that midpoint from the origin is clearly $\frac{\sqrt{2}}{2}$, the length of C being $\sqrt{2}$.

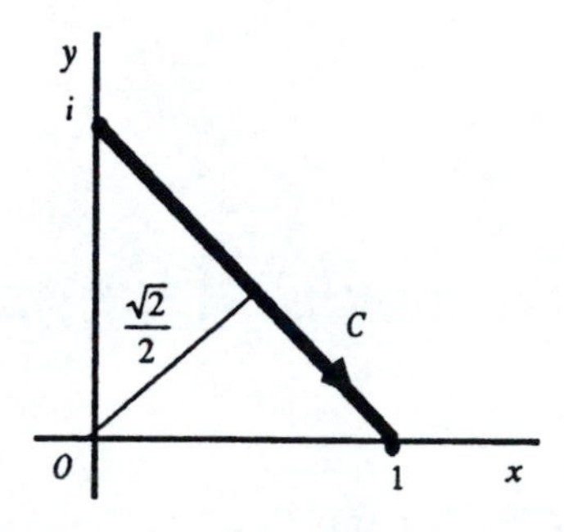

Hence if z is any point on C, $|z| \ge \frac{\sqrt{2}}{2}$. This means that, for such a point $\left|\frac{1}{z^4}\right| = \frac{1}{|z|^4} \le 4$. Consequently, by taking $M = 4$ and $L = \sqrt{2}$, we have

$$\left|\int_C \frac{dz}{z^4}\right| \le ML = 4\sqrt{2}.$$

3. The contour C is the closed triangular path shown below.

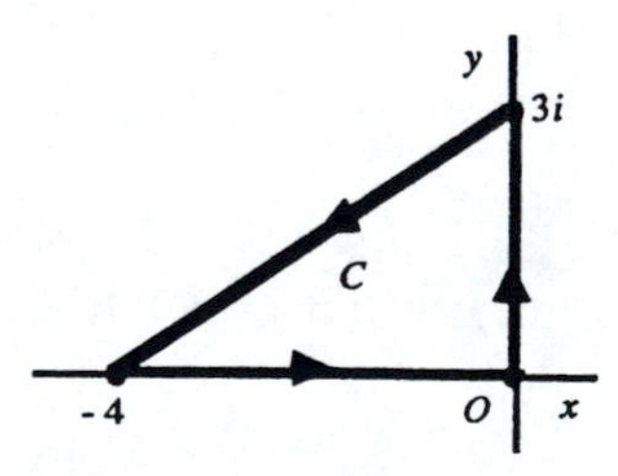

To find an upper bound for $\left|\int_C (e^z - \bar{z})dz\right|$, we let z be a point on C and observe that

$$|e^z - \bar{z}| \le |e^z| + |\bar{z}| = e^x + \sqrt{x^2 + y^2}.$$

But $e^x \le 1$ since $x \le 0$, and the distance $\sqrt{x^2+y^2}$ of the point z from the origin is always less than or equal to 4. Thus $|e^z - \bar{z}| \le 5$ when z is on C. The length of C is evidently 12. Hence, by writing $M = 5$ and $L = 12$, we have

$$\left|\int_C (e^z - \bar{z})dz\right| \le ML = 60.$$

4. Note that if $|z| = R$ $(R > 2)$, then

$$|2z^2 - 1| \le 2|z|^2 + 1 = 2R^2 + 1$$

and

$$|z^4 + 5z^2 + 4| = |z^2 + 1||z^2 + 4| \ge \left||z|^2 - 1\right|\left||z|^2 - 4\right| = (R^2 - 1)(R^2 - 4).$$

Thus

$$\left|\frac{2z^2 - 1}{z^4 + 5z^2 + 4}\right| = \frac{|2z^2 - 1|}{|z^4 + 5z^2 + 4|} \le \frac{2R^2 + 1}{(R^2 - 1)(R^2 - 4)}$$

when $|z| = R$ $(R > 2)$. Since the length of C_R is πR, then,

$$\left|\int_{C_R} \frac{2z^2 - 1}{z^4 + 5z^2 + 4} dz\right| \le \frac{\pi R(2R^2 + 1)}{(R^2 - 1)(R^2 - 4)} = \frac{\frac{\pi}{R}\left(2 + \frac{1}{R^2}\right)}{\left(1 - \frac{1}{R^2}\right)\left(1 - \frac{4}{R^2}\right)};$$

and it is clear that the value of the integral tends to zero as R tends to infinity.

5. Here C_R is the positively oriented circle $|z| = R\,(R > 1)$. If z is a point on C_R, then

$$\left|\frac{\operatorname{Log} z}{z^2}\right| = \frac{|\ln R + i\Theta|}{R^2} \le \frac{\ln R + |\Theta|}{R^2} \le \frac{\pi + \ln R}{R^2},$$

since $-\pi < \Theta \le \pi$. The length of C_R is, of course, $2\pi R$. Consequently, by taking

$$M = \frac{\pi + \ln R}{R^2} \quad \text{and} \quad L = 2\pi R,$$

we see that

$$\left|\int_{C_R} \frac{\operatorname{Log} z}{z^2}\, dz\right| \le ML = 2\pi\left(\frac{\pi + \ln R}{R}\right).$$

Since

$$\lim_{R\to\infty} \frac{\pi + \ln R}{R} = \lim_{R\to\infty} \frac{1/R}{1} = 0,$$

it follows that

$$\lim_{R\to\infty} \int_{C_R} \frac{\operatorname{Log} z}{z^2}\, dz = 0.$$

6. Let C_ρ be the positively oriented circle $|z| = \rho\ (0 < \rho < 1)$, shown in the figure below, and suppose that $f(z)$ is analytic in the disk $|z| \le 1$.

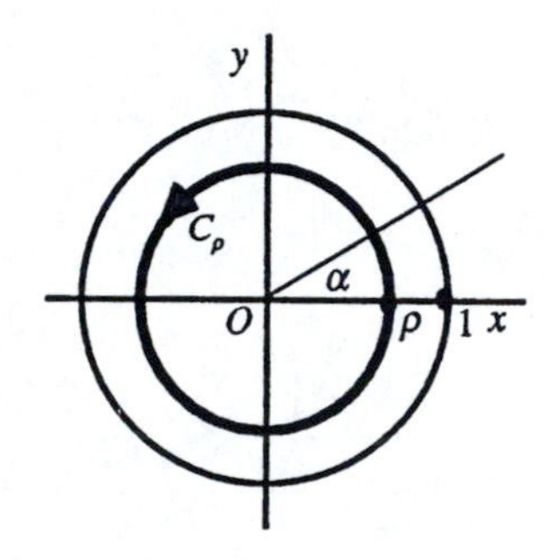

We let $z^{-1/2}$ represent any particular branch

$$z^{-1/2} = \exp\left(-\frac{1}{2}\log z\right) = \exp\left[-\frac{1}{2}(\ln r + i\theta)\right] = \frac{1}{\sqrt{r}}\exp\left(-i\frac{\theta}{2}\right) \qquad (r > 0, \alpha < \theta < \alpha + 2\pi)$$

of the power function here; and we note that, since $f(z)$ is continuous on the *closed bounded* disk $|z| \le 1$, there is a nonnegative constant M such that $|f(z)| \le M$ for each point z in that disk. We are asked to find an upper bound for $\left|\int_{C_\rho} z^{-1/2} f(z)dz\right|$. To do this, we observe that if z is a point on C_ρ,

$$\left|z^{-1/2} f(z)\right| = \left|z^{-1/2}\right| |f(z)| \le \frac{M}{\sqrt{\rho}}.$$

Since the length of the path C_ρ is $2\pi\rho$, we may conclude that

$$\left|\int_{C_\rho} z^{-1/2} f(z)dz\right| \le \frac{M}{\sqrt{\rho}} 2\pi\rho = 2\pi M\sqrt{\rho}.$$

Note that, inasmuch as M is independent of ρ, it follows that

$$\lim_{\rho\to 0} \int_{C_\rho} z^{-1/2} f(z)dz = 0.$$

SECTION 43

1. The function z^n $(n = 0,1,2,\ldots)$ has the antiderivative $z^{n+1}/(n+1)$ everywhere in the finite plane. Consequently, for any contour C from a point z_1 to a point z_2,

$$\int_C z^n dz = \int_{z_1}^{z_2} z^n dz = \left.\frac{z^{n+1}}{n+1}\right]_{z_1}^{z_2} = \frac{z_2^{n+1}}{n+1} - \frac{z_1^{n+1}}{n+1} = \frac{1}{n+1}\left(z_2^{n+1} - z_1^{n+1}\right).$$

2. *(a)* $\displaystyle\int_i^{i/2} e^{\pi z}\,dz = \left.\frac{e^{\pi z}}{\pi}\right]_i^{i/2} = \frac{e^{i\pi/2} - e^{i\pi}}{\pi} = \frac{i+1}{\pi} = \frac{1+i}{\pi}.$

(b) $\displaystyle\int_0^{\pi+2i} \cos\left(\frac{z}{2}\right) dz = 2\sin\left(\frac{z}{2}\right)\Bigg]_0^{\pi+2i} = 2\sin\left(\frac{\pi}{2}+i\right) = 2\frac{e^{i\left(\frac{\pi}{2}+i\right)} - e^{-i\left(\frac{\pi}{2}+i\right)}}{2i} = -i\left(e^{i\pi/2}e^{-1} - e^{-i\pi/2}e\right)$

$$= -i\left(\frac{i}{e} + ie\right) = \frac{1}{e} + e = e + \frac{1}{e}.$$

(c) $\displaystyle\int_1^3 (z-2)^3\,dz = \left.\frac{(z-2)^4}{4}\right]_1^3 = \frac{1}{4} - \frac{1}{4} = 0.$

3. Note the function $(z-z_0)^{n-1}$ $(n = \pm 1, \pm 2,\ldots)$ always has an antiderivative in any domain that does not contain the point $z = z_0$. So, by the theorem in Sec. 42,

$$\int_{C_0} (z-z_0)^{n-1} dz = 0$$

for any closed contour C_0 that does not pass through z_0.

5. Let C denote any contour from $z = -1$ to $z = 1$ that, except for its end points, lies above the real axis. This exercise asks us to evaluate the integral

$$I = \int_{-1}^{1} z^i dz,$$

where z^i denotes the principal branch

$$z^i = \exp(i\,\mathrm{Log} z) \qquad (|z| > 0, -\pi < \mathrm{Arg}\, z < \pi).$$

An antiderivative of this branch *cannot* be used since the branch is not even defined at $z=-1$. But the integrand can be replaced by the branch

$$z^i = \exp(i\log z) \qquad \left(|z|>0, -\frac{\pi}{2}<\arg z<\frac{3\pi}{2}\right)$$

since it agrees with the integrand along C. Using an antiderivative of this new branch, we can now write

$$I = \frac{z^{i+1}}{i+1}\Bigg]_{-1}^{1} = \frac{1}{i+1}\left[(1)^{i+1}-(-1)^{i+1}\right] = \frac{1}{i+1}\left[e^{(i+1)\log 1} - e^{(1+1)\log(-1)}\right]$$

$$= \frac{1}{i+1}\left[e^{(i+1)(\ln 1+i0)} - e^{(i+1)(\ln 1+i\pi)}\right] = \frac{1}{i+1}\left(1-e^{-\pi}e^{i\pi}\right) = \frac{1+e^{-\pi}}{1+i}\cdot\frac{1-i}{1-i}$$

$$= \frac{1+e^{-\pi}}{2}(1-i).$$

SECTION 46

2. The contours C_1 and C_2 are as shown in the figure below.

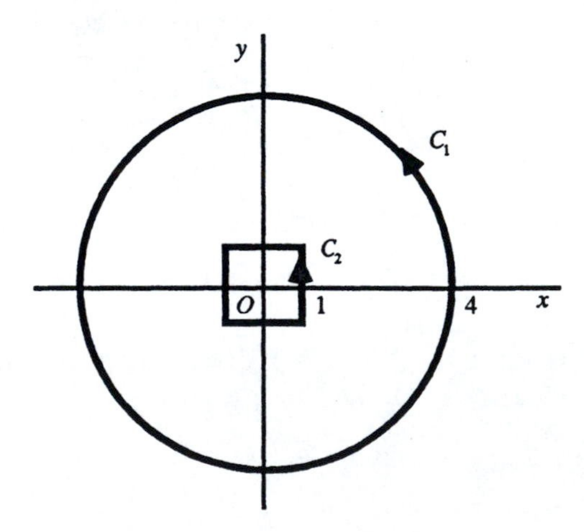

In each of the cases below, the singularities of the integrand lie outside C_1 or inside C_2; and so the integrand is analytic on the contours and between them. Consequently,

$$\int_{C_1} f(z)\,dz = \int_{C_2} f(z)\,dz.$$

(a) When $f(z)=\dfrac{1}{3z^2+1}$, the singularities are the points $z=\pm\dfrac{1}{\sqrt{3}}i$.

(b) When $f(z)=\dfrac{z+2}{\sin(z/2)}$, the singularities are at $z=2n\pi\ (n=0,\pm1,\pm2,\ldots)$.

(c) When $f(z)=\dfrac{z}{1-e^z}$, the singularities are at $z=2n\pi i\ (n=0,\pm1,\pm2,\ldots)$.

4. *(a)* In order to derive the integration formula in question, we integrate the function e^{-z^2} around the closed rectangular path shown below.

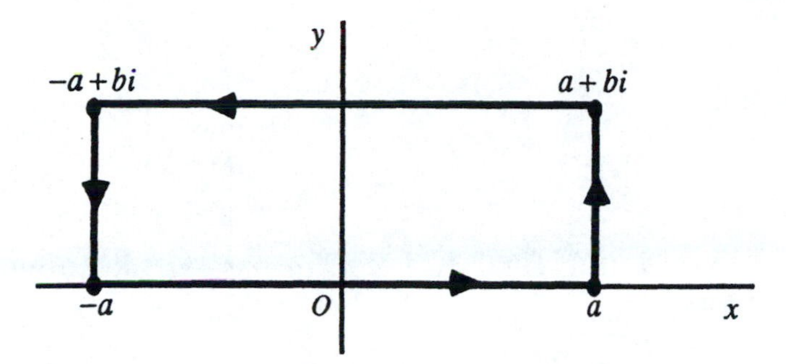

Since the lower horizontal leg is represented by $z=x\ (-a\le x\le a)$, the integral of e^{-z^2} along that leg is

$$\int_{-a}^{a} e^{-x^2}\,dx = 2\int_{0}^{a} e^{-x^2}\,dx.$$

Since the opposite direction of the upper horizontal leg has parametric representation $z=x+bi\ (-a\le x\le a)$, the integral of e^{-z^2} along the upper leg is

$$-\int_{-a}^{a} e^{-(x+bi)^2}\,dx = -e^{b^2}\int_{-a}^{a} e^{-x^2}e^{-i2bx}\,dx = -e^{b^2}\int_{-a}^{a} e^{-x^2}\cos 2bx\,dx + ie^{b^2}\int_{-a}^{a} e^{-x^2}\sin 2bx\,dx,$$

or simply

$$-2e^{b^2}\int_{0}^{a} e^{-x^2}\cos 2bx\,dx.$$

Since the right-hand vertical leg is represented by $z=a+iy\ (0\le y\le b)$, the integral of e^{-z^2} along it is

$$\int_{0}^{b} e^{-(a+iy)^2}\,i\,dy = ie^{-a^2}\int_{0}^{b} e^{y^2}e^{-i2ay}\,dy.$$

Finally, since the opposite direction of the left-hand vertical leg has the representation $z=-a+iy$ $(0 \le y \le b)$, the integral of e^{-z^2} along that vertical leg is

$$-\int_0^b e^{-(-a+iy)^2} i\,dy = -ie^{-a^2}\int_0^b e^{y^2} e^{i2ay}\,dy.$$

According to the Cauchy-Goursat theorem, then,

$$2\int_0^a e^{-x^2}dx - 2e^{b^2}\int_0^a e^{-x^2}\cos 2bx\,dx + ie^{-a^2}\int_0^b e^{y^2}e^{-i2ay}dy - ie^{-a^2}\int_0^b e^{y^2}e^{i2ay}dy = 0;$$

and this reduces to

$$\int_0^a e^{-x^2}\cos 2bx\,dx = e^{-b^2}\int_0^a e^{-x^2}dx + e^{-(a^2+b^2)}\int_0^b e^{y^2}\sin 2ay\,dy.$$

(b) We now let $a \to \infty$ in the final equation in part *(a)*, keeping in mind the known integration formula

$$\int_0^\infty e^{-x^2}dx = \frac{\sqrt{\pi}}{2}$$

and the fact that

$$\left| e^{-(a^2+b^2)}\int_0^b e^{y^2}\sin 2ay\,dy \right| \le e^{-(a^2+b^2)}\int_0^b e^{y^2}\,dy \to 0 \text{ as } a \to \infty.$$

The result is

$$\int_0^\infty e^{-x^2}\cos 2bx\,dx = \frac{\sqrt{\pi}}{2}e^{-b^2} \qquad (b>0).$$

6. We let C denote the entire boundary of the semicircular region appearing below. It is made up of the leg C_1 from the origin to the point $z=1$, the semicircular arc C_2 that is shown, and the leg C_3 from $z=-1$ to the origin. Thus $C = C_1 + C_2 + C_3$.

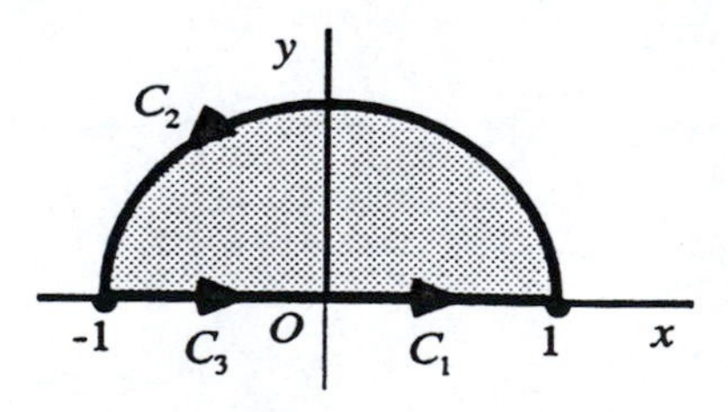

We also let $f(z)$ be a continuous function that is defined on this closed semicircular region by writing $f(0)=0$ and using the branch

$$f(z)=\sqrt{r}e^{i\theta/2} \qquad \left(r>0, -\frac{\pi}{2}<\theta<\frac{3\pi}{2}\right)$$

of the multiple-valued function $z^{1/2}$. The problem here is to evaluate the integral of $f(z)$ around C by evaluating the integrals along the individual paths $C_1, C_2,$ and C_3 and then adding the results. In each case, we write a parametric representation for the path (or a related one) and then use it to evaluate the integral along the particular path.

(i) C_1: $z=re^{i0}$ $(0\le r\le 1)$. Then

$$\int_{C_1} f(z)\,dz=\int_0^1 \sqrt{r}\cdot 1\,dr=\left[\frac{2}{3}r^{3/2}\right]_0^1=\frac{2}{3}.$$

(ii) C_2: $z=1\cdot e^{i\theta}$ $(0\le\theta\le\pi)$. Then

$$\int_{C_2} f(z)\,dz=\int_0^\pi e^{i\theta/2}\cdot ie^{i\theta}d\theta=i\int_0^\pi e^{i3\theta/2}d\theta=i\left[\frac{2}{3i}e^{i3\theta/2}\right]_0^\pi=\frac{2}{3}(-i-1)=-\frac{2}{3}(1+i).$$

(iii) $-C_3$: $z=re^{i\pi}$ $(0\le r\le 1)$. Then

$$\int_{C_3} f(z)\,dz=-\int_{-C_3} f(z)\,dz=-\int_0^1 \sqrt{r}e^{i\pi/2}(-1)dr=i\int_0^1\sqrt{r}dr=i\left[\frac{2}{3}r^{3/2}\right]_0^1=\frac{2}{3}i.$$

The desired result is

$$\int_C f(z)\,dz=\int_{C_1} f(z)\,dz+\int_{C_2} f(z)\,dz+\int_{C_3} f(z)\,dz=\frac{2}{3}-\frac{2}{3}(1+i)+\frac{2}{3}i=0.$$

The Cauchy-Goursat theorem does not apply since $f(z)$ is not analytic at the origin, or even defined on the negative imaginary axis.

SECTION 48

1. In this problem, we let C denote the square contour shown in the figure below.

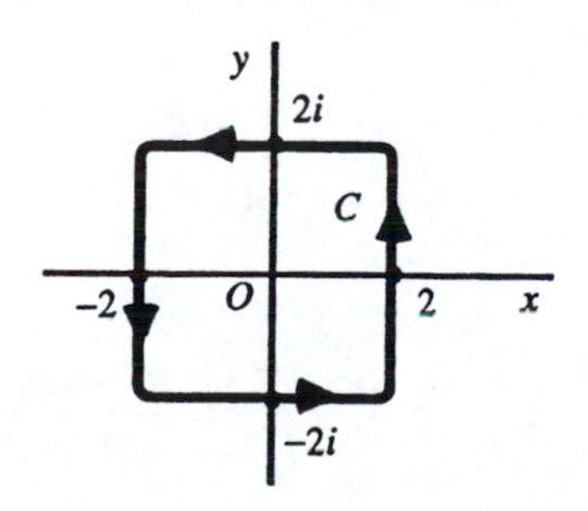

(a) $\int_C \frac{e^{-z}\,dz}{z-(\pi i/2)} = 2\pi i\left[e^{-z}\right]_{z=\pi i/2} = 2\pi i(-i) = 2\pi.$

(b) $\int_C \frac{\cos z}{z(z^2+8)}\,dz = \int_C \frac{(\cos z)/(z^2+8)}{z-0}\,dz = 2\pi i\left[\frac{\cos z}{z^2+8}\right]_{z=0} = 2\pi i\left(\frac{1}{8}\right) = \frac{\pi i}{4}.$

(c) $\int_C \frac{z\,dz}{2z+1} = \int_C \frac{z/2}{z-(-1/2)}\,dz = 2\pi i\left[\frac{z}{2}\right]_{z=-1/2} = 2\pi i\left(-\frac{1}{4}\right) = -\frac{\pi i}{2}.$

(d) $\int_C \frac{\cosh z}{z^4}\,dz = \int_C \frac{\cosh z}{(z-0)^{3+1}}\,dz = \frac{2\pi i}{3!}\left[\frac{d^3}{dz^3}\cosh z\right]_{z=0} = \frac{\pi i}{3}(0) = 0.$

(e) $\int_C \frac{\tan(z/2)}{(z-x_0)^2}\,dz = \int_C \frac{\tan(z/2)}{(z-x_0)^{1+1}}\,dz = \frac{2\pi i}{1!}\left[\frac{d}{dz}\tan\left(\frac{z}{2}\right)\right]_{z=x_0}$

$$= 2\pi i\left(\frac{1}{2}\sec^2\frac{x_0}{2}\right) = i\pi\sec^2\left(\frac{x_0}{2}\right) \text{ when } -2 < x_0 < 2.$$

2. Let C denote the positively oriented circle $|z-i| = 2$, shown below.

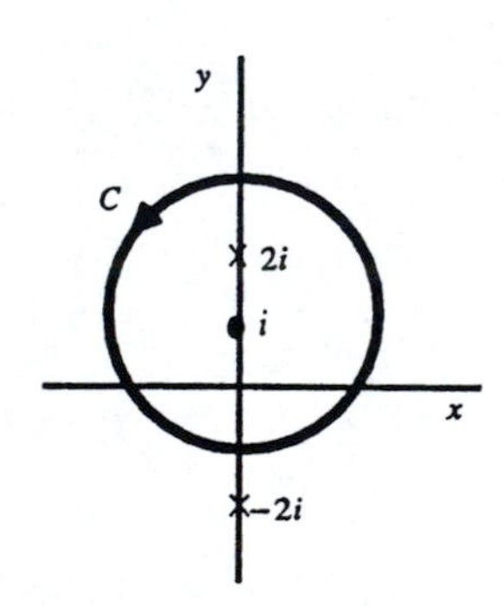

(a) The Cauchy integral formula enables us to write

$$\int_C \frac{dz}{z^2+4} = \int_C \frac{dz}{(z-2i)(z+2i)} = \int_C \frac{1/(z+2i)}{z-2i}\,dz = 2\pi i\left(\frac{1}{z+2i}\right)_{z=2i} = 2\pi i\left(\frac{1}{4i}\right) = \frac{\pi}{2}.$$

(b) Applying the extended form of the Cauchy integral formula, we have

$$\int_C \frac{dz}{(z^2+4)^2} = \int_C \frac{dz}{(z-2i)^2(z+2i)^2} = \int_C \frac{1/(z+2i)^2}{(z-2i)^{1+1}}\,dz = \frac{2\pi i}{1!}\left[\frac{d}{dz}\frac{1}{(z+2i)^2}\right]_{z=2i}$$

$$= 2\pi i\left[\frac{-2}{(z+2i)^3}\right]_{z=2i} = \frac{-4\pi i}{(4i)^3} = \frac{-4\pi i}{-(16)(4)i} = \frac{\pi}{16}.$$

3. Let C be the positively oriented circle $|z|=3$, and consider the function

$$g(w) = \int_C \frac{2z^2 - z - 2}{z - w}\,dz \qquad (|w| \neq 3).$$

We wish to find $g(w)$ when $w = 2$ and when $|w| > 3$ (see the figure below).

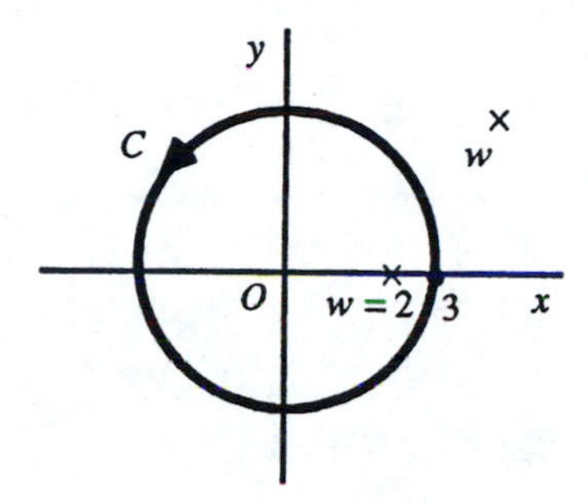

We observe that

$$g(2) = \int_C \frac{2z^2 - z - 2}{z - 2}\,dz = 2\pi i\left[2z^2 - z - 2\right]_{z=2} = 2\pi i(4) = 8\pi i.$$

On the other hand, when $|w| > 3$, the Cauchy-Goursat theorem tells us that $g(w) = 0$.

5. Suppose that a function f is analytic inside and on a simple closed contour C and that z_0 is not on C. If z_0 is inside C, then

$$\int_C \frac{f'(z)\,dz}{z - z_0} = 2\pi i f'(z_0) \quad \text{and} \quad \int_C \frac{f(z)\,dz}{(z - z_0)^2} = \int_C \frac{f(z)\,dz}{(z - z_0)^{1+1}} = \frac{2\pi i}{1!} f'(z_0).$$

Thus

$$\int_C \frac{f'(z)\,dz}{z - z_0} = \int_C \frac{f(z)\,dz}{(z - z_0)^2}.$$

The Cauchy-Goursat theorem tells us that this last equation is also valid when z_0 is exterior to C, each side of the equation being 0.

7. Let C be the unit circle $z = e^{i\theta}$ $(-\pi \le \theta \le \pi)$, and let a denote any real constant. The Cauchy integral formula reveals that

$$\int_C \frac{e^{az}}{z}\,dz = \int_C \frac{e^{az}}{z - 0}\,dz = 2\pi i\left[e^{az}\right]_{z=0} = 2\pi i.$$

On the other hand, the stated parametric representation for C gives us

$$\int_C \frac{e^{az}}{z}dz = \int_{-\pi}^{\pi} \frac{\exp(ae^{i\theta})}{e^{i\theta}} ie^{i\theta}d\theta = i\int_{-\pi}^{\pi} \exp[a(\cos\theta + i\sin\theta)]d\theta$$

$$= i\int_{-\pi}^{\pi} e^{a\cos\theta}e^{ia\sin\theta}d\theta = i\int_{-\pi}^{\pi} e^{a\cos\theta}[\cos(a\sin\theta) + i\sin(a\sin\theta)]d\theta$$

$$= -\int_{-\pi}^{\pi} e^{a\cos\theta}\sin(a\sin\theta)d\theta + i\int_{-\pi}^{\pi} e^{a\cos\theta}\cos(a\sin\theta)d\theta.$$

Equating these two different expressions for the integral $\int_C \frac{e^{az}}{z}dz$, we have

$$-\int_{-\pi}^{\pi} e^{a\cos\theta}\sin(a\sin\theta)d\theta + i\int_{-\pi}^{\pi} e^{a\cos\theta}\cos(a\sin\theta)d\theta = 2\pi i.$$

Then, by equating the imaginary parts on each side of this last equation, we see that

$$\int_{-\pi}^{\pi} e^{a\cos\theta}\cos(a\sin\theta)d\theta = 2\pi;$$

and, since the integrand here is even,

$$\int_{0}^{\pi} e^{a\cos\theta}\cos(a\sin\theta)d\theta = \pi.$$

8. *(a)* The binomial formula enables us to write

$$P_n(z) = \frac{1}{n!2^n}\frac{d^n}{dz^n}(z^2-1)^n = \frac{1}{n!2^n}\frac{d^n}{dz^n}\sum_{k=0}^{n}\binom{n}{k}z^{2n-2k}(-1)^k.$$

We note that the highest power of z appearing under the derivative is z^{2n}, and differentiating it n times brings it down to z^n. So $P_n(z)$ is a polynomial of degree n.

(b) We let C denote any positively oriented simple closed contour surrounding a fized point z. The Cauchy integral formula for derivatives tells us that

$$\frac{d^n}{dz^n}(z^2-1)^n = \frac{n!}{2\pi i}\int_C \frac{(s^2-1)^n}{(s-z)^{n+1}}ds \qquad (n = 0,1,2,\ldots).$$

Hence the polynomials $P_n(z)$ in part *(a)* can be written

$$P_n(z) = \frac{1}{2^{n+1}\pi i}\int_C \frac{(s^2-1)^n}{(s-z)^{n+1}}ds \qquad (n = 0,1,2,\ldots).$$

(c) Note that

$$\frac{(s^2-1)^n}{(s-1)^{n+1}} = \frac{(s-1)^n(s+1)^n}{(s-1)^{n+1}} = \frac{(s+1)^n}{s-1}.$$

Referring to the final result in part *(b)*, then, we have

$$P_n(1) = \frac{1}{2^{n+1}\pi i}\int_C \frac{(s^2-1)^n}{(s-1)^{n+1}}ds = \frac{1}{2^n}\frac{1}{2\pi i}\int_C \frac{(s+1)^n}{s-1}ds = \frac{1}{2^n}2^n = 1 \qquad (n = 0,1,2,\ldots).$$

Also, since

$$\frac{(s^2-1)^n}{(s+1)^{n+1}} = \frac{(s-1)^n(s+1)^n}{(s+1)^{n+1}} = \frac{(s-1)^n}{s+1},$$

we have

$$P_n(-1) = \frac{1}{2^{n+1}\pi i}\int_C \frac{(s^2-1)^n}{(s+1)^{n+1}}ds = \frac{1}{2^n}\frac{1}{2\pi i}\int_C \frac{(s-1)^n}{s+1}ds = \frac{1}{2^n}(-2)^n = (-1)^n \quad (n = 0,1,2,\ldots).$$

9. We are asked to show that

$$f''(z) = \frac{1}{\pi i}\int_C \frac{f(s)\,ds}{(s-z)^3}.$$

(a) In view of the expression for $f'(z)$ in the lemma,

$$\begin{aligned}\frac{f'(z+\Delta z)-f'(z)}{\Delta z} &= \frac{1}{2\pi i}\int_C\left[\frac{1}{(s-z-\Delta z)^2}-\frac{1}{(s-z)^2}\right]\frac{f(s)ds}{\Delta z}\\ &= \frac{1}{2\pi i}\int_C \frac{2(s-z)-\Delta z}{(s-z-\Delta z)^2(s-z)^2}f(s)ds.\end{aligned}$$

Then

$$\begin{aligned}\frac{f'(z+\Delta z)-f'(z)}{\Delta z}-\frac{1}{\pi i}\int_C\frac{f(s)ds}{(s-z)^3} &= \frac{1}{2\pi i}\int_C\left[\frac{2(s-z)-\Delta z}{(s-z-\Delta z)^2(s-z)^2}-\frac{2}{(s-z)^3}\right]f(s)ds\\ &= \frac{1}{2\pi i}\int_C\frac{3(s-z)\Delta z-2(\Delta z)^2}{(s-z-\Delta z)^2(s-z)^3}f(s)ds.\end{aligned}$$

(b) We must show that

$$\left|\int_C \frac{3(s-z)\Delta z - 2(\Delta z)^2}{(s-z-\Delta z)^2(s-z)^3} f(s)ds\right| \le \frac{(3D|\Delta z| + 2|\Delta z|^2)M}{(d-|\Delta z|)^2 d^3} L.$$

Now D, d, M, and L are as in the statement of the exercise in the text. The triangle inequality tells us that

$$|3(s-z)\Delta z - 2(\Delta z)^2| \le 3|s-z||\Delta z| + 2|\Delta z|^2 \le 3D|\Delta z| + 2|\Delta z|^2.$$

Also, we know from the verification of the expression for $f'(z)$ in the lemma that $|s-z-\Delta z| \ge d - |\Delta z| > 0$; and this means that

$$|(s-z-\Delta z)^2(s-z)^3| \ge (d-|\Delta z|)^2 d^3 > 0.$$

This gives the desired inequality.

(c) If we let Δz tend to 0 in the inequality obtained in part *(b)* we find that

$$\lim_{\Delta z \to 0} \frac{1}{2\pi i}\int_C \frac{3(s-z)\Delta z - 2(\Delta z)^2}{(s-z-\Delta z)^2(s-z)^3} f(s)ds = 0.$$

This, together with the result in part *(a)*, yields the desided expression for $f''(z)$.

Chapter 5

SECTION 52

1. We are asked to show in two ways that the sequence

$$z_n = -2 + i\frac{(-1)^n}{n^2} \qquad (n = 1,2,\ldots)$$

converges to -2. One way is to note that the two sequences

$$x_n = -2 \quad \text{and} \quad y_n = \frac{(-1)^n}{n^2} \qquad (n = 1,2,\ldots)$$

of real numbers converge to -2 and 0, respectively, and then to apply the theorem in Sec. 51. Another way is to observe that $|z_n - (-2)| = \frac{1}{n^2}$. Thus for each $\varepsilon > 0$,

$$|z_n - (-2)| < \varepsilon \quad \text{whenever} \quad n > n_0,$$

where n_0 is any positive integer such that $n_0 \geq \frac{1}{\sqrt{\varepsilon}}$.

2. Observe that if $z_n = -2 + i\frac{(-1)^n}{n^2}$ $(n = 1,2,\ldots)$, then

$$r_n = |z_n| = \sqrt{4 + \frac{1}{n^4}} \to 2.$$

But, since

$$\Theta_{2n} = \operatorname{Arg} z_{2n} \to \pi \quad \text{and} \quad \Theta_{2n-1} = \operatorname{Arg} z_{2n-1} \to -\pi \qquad (n = 1,2,\ldots),$$

the sequence Θ_n $(n = 1,2,\ldots)$ does not converge.

3. Suppose that $\lim_{n\to\infty} z_n = z$. That is, for each $\varepsilon > 0$, there is a positive integer n_0 such that $|z_n - z| < \varepsilon$ whenever $n > n_0$. In view of the inequality (see Sec. 4)

$$|z_n - z| \geq ||z_n| - |z||,$$

it follows that $||z_n| - |z|| < \varepsilon$ whenever $n > n_0$. That is, $\lim_{n\to\infty} |z_n| = |z|$.

4. The summation formula found in the example in Sec. 52 can be written

$$\sum_{n=1}^{\infty} z^n = \frac{z}{1-z} \quad \text{when} \quad |z|<1.$$

If we put $z = re^{i\theta}$, where $0<r<1$, the left-hand side becomes

$$\sum_{n=1}^{\infty} (re^{i\theta})^n = \sum_{n=1}^{\infty} r^n e^{in\theta} = \sum_{n=1}^{\infty} r^n \cos n\theta + i\sum_{n=1}^{\infty} r^n \sin n\theta;$$

and the right-hand side takes the form

$$\frac{re^{i\theta}}{1-re^{i\theta}} \cdot \frac{1-re^{-i\theta}}{1-re^{-i\theta}} = \frac{re^{i\theta}-r^2}{1-r(e^{i\theta}+e^{-i\theta})+r^2} = \frac{r\cos\theta - r^2 + ir\sin\theta}{1-2r\cos\theta+r^2}.$$

Thus

$$\sum_{n=1}^{\infty} r^n \cos n\theta + i\sum_{n=1}^{\infty} r^n \sin n\theta = \frac{r\cos\theta - r^2}{1-2r\cos\theta+r^2} + i\frac{r\sin\theta}{1-2r\cos\theta+r^2}.$$

Equating the real parts on each side here and then the imaginary parts, we arrive at the summation formulas

$$\sum_{n=1}^{\infty} r^n \cos n\theta = \frac{r\cos\theta - r^2}{1-2r\cos\theta+r^2} \quad \text{and} \quad \sum_{n=1}^{\infty} r^n \sin n\theta = \frac{r\sin\theta}{1-2r\cos\theta+r^2},$$

where $0<r<1$. These formulas clearly hold when $r=0$ too.

6. Suppose that $\sum_{n=1}^{\infty} z_n = S$. To show that $\sum_{n=1}^{\infty} \overline{z}_n = \overline{S}$, we write $z_n = x_n + iy_n$, $S = X + iY$ and appeal to the theorem in Sec. 52. First of all, we note that

$$\sum_{n=1}^{\infty} x_n = X \quad \text{and} \quad \sum_{n=1}^{\infty} y_n = Y.$$

Then, since $\sum_{n=1}^{\infty} (-y_n) = -Y$, it follows that

$$\sum_{n=1}^{\infty} \overline{z}_n = \sum_{n=1}^{\infty} (x_n - iy_n) = \sum_{n=1}^{\infty} [x_n + i(-y_n)] = X - iY = \overline{S}.$$

8. Suppose that $\sum_{n=1}^{\infty} z_n = S$ and $\sum_{n=1}^{\infty} w_n = T$. In order to use the theorem in Sec. 52, we write

$$z_n = x_n + iy_n, \quad S = X + iY \quad \text{and} \quad w_n = u_n + iv_n, \quad T = U + iV.$$

Now

$$\sum_{n=1}^{\infty} x_n = X, \quad \sum_{n=1}^{\infty} y_n = Y \quad \text{and} \quad \sum_{n=1}^{\infty} u_n = U, \quad \sum_{n=1}^{\infty} v_n = V.$$

Since

$$\sum_{n=1}^{\infty} (x_n + u_n) = X + U \quad \text{and} \quad \sum_{n=1}^{\infty} (y_n + v_n) = Y + V,$$

it follows that

$$\sum_{n=1}^{\infty} [(x_n + u_n) + i(y_n + v_n)] = X + U + i(Y + V).$$

That is,

$$\sum_{n=1}^{\infty} [(x_n + iy_n) + (u_n + iv_n)] = X + iY + (U + iV),$$

or

$$\sum_{n=1}^{\infty} (z_n + w_n) = S + T.$$

SECTION 54

1. Replace z by z^2 in the known series

$$\cosh z = \sum_{n-0}^{\infty} \frac{z^{2n}}{(2n)!} \qquad (|z| < \infty)$$

to get

$$\cosh(z^2) = \sum_{n=0}^{\infty} \frac{z^{4n}}{(2n)!} \qquad (|z| < \infty).$$

Then, multiplying through this last equation by z, we have the desired result:

$$z\cosh(z^2) = \sum_{n=0}^{\infty} \frac{z^{4n+1}}{(2n)!} \qquad (|z| < \infty).$$

2. *(b)* Replacing z by $z-1$ in the known expansion

$$e^z = \sum_{n=0}^{\infty} \frac{z^n}{n!} \qquad (|z| < \infty),$$

we have

$$e^{z-1} = \sum_{n=0}^{\infty} \frac{(z-1)^n}{n!} \qquad (|z| < \infty).$$

So

$$e^z = e^{z-1}e = e\sum_{n=0}^{\infty} \frac{(z-1)^n}{n!} \qquad (|z| < \infty).$$

3. We want to find the Maclaurin series for the function

$$f(z) = \frac{z}{z^4+9} = \frac{z}{9} \cdot \frac{1}{1+(z^4/9)}.$$

To do this, we first replace z by $-(z^4/9)$ in the known expansion

$$\frac{1}{1-z} = \sum_{n=0}^{\infty} z^n \qquad (|z| < 1),$$

as well as its condition of validity, to get

$$\frac{1}{1+(z^4/9)} = \sum_{n=0}^{\infty} \frac{(-1)^n}{3^{2n}} z^{4n} \qquad (|z| < \sqrt{3}).$$

Then, if we multiply through this last equation by $\frac{z}{9}$, we have the desired expansion:

$$f(z) = \sum_{n=0}^{\infty} \frac{(-1)^n}{3^{2n+2}} z^{4n+1} \qquad (|z| < \sqrt{3}).$$

6. Replacing z by z^2 in the representation

$$\sin z = \sum_{n=0}^{\infty} (-1)^n \frac{z^{2n+1}}{(2n+1)!} \qquad (|z| < \infty),$$

we have

$$\sin(z^2) = \sum_{n=0}^{\infty} (-1)^n \frac{z^{4n+2}}{(2n+1)!} \qquad (|z| < \infty).$$

Since the coefficient of z^n in the Maclaurin series for a function $f(z)$ is $f^{(n)}(0)/n!$, this shows that

$$f^{(4n)}(0)=0 \quad \text{and} \quad f^{(2n+1)}(0)=0 \qquad (n=0,1,2,\ldots).$$

7. The function $\dfrac{1}{1-z}$ has a singularity at $z=1$. So the Taylor series about $z=i$ is valid when $|z-i|<\sqrt{2}$, as indicated in the figure below.

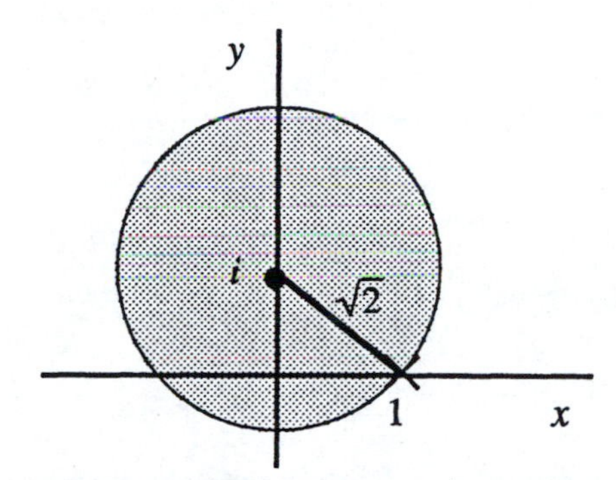

To find the series, we start by writing

$$\frac{1}{1-z}=\frac{1}{(1-i)-(z-i)}=\frac{1}{1-i}\cdot\frac{1}{1-(z-i)/(1-i)}.$$

This suggests that we replace z by $(z-i)/(1-i)$ in the known expansion

$$\frac{1}{1-z}=\sum_{n=0}^{\infty} z^n \qquad (|z|<1)$$

and then multiply through by $\dfrac{1}{1-i}$. The desired Taylor series is then obtained:

$$\frac{1}{1-z}=\sum_{n=0}^{\infty}\frac{(z-i)^n}{(1-i)^{n+1}} \qquad (|z-i|<\sqrt{2}).$$

9 The identity $\sinh(z+\pi i)=-\sinh z$ and the periodicity of $\sinh z$, with period $2\pi i$, tell us that

$$\sinh z=-\sinh(z+\pi i)=-\sinh(z-\pi i).$$

So, if we replace z by $z-\pi i$ in the known representation

$$\sinh z=\sum_{n=0}^{\infty}\frac{z^{2n+1}}{(2n+1)!} \qquad (|z|<\infty)$$

and then multiply through by -1, we find that

$$\sinh z = -\sum_{n=0}^{\infty} \frac{(z-\pi i)^{2n+1}}{(2n+1)!} \qquad (|z-\pi i| < \infty).$$

13. Suppose that $0 < |z| < 4$. Then $0 < |z/4| < 1$, and we can use the known expansion

$$\frac{1}{1-z} = \sum_{n=0}^{\infty} z^n \qquad (|z| < 1).$$

To be specific, when $0 < |z| < 4$,

$$\frac{1}{4z-z^2} = \frac{1}{4z} \cdot \frac{1}{1-\frac{z}{4}} = \frac{1}{4z} \sum_{n=0}^{\infty} \left(\frac{z}{4}\right)^n = \sum_{n=0}^{\infty} \frac{z^{n-1}}{4^{n+1}} = \frac{1}{4z} + \sum_{n=1}^{\infty} \frac{z^{n-1}}{4^{n+1}} = \frac{1}{4z} + \sum_{n=0}^{\infty} \frac{z^n}{4^{n+2}}.$$

SECTION 56

1. We may use the expansion

$$\sin z = \sum_{n=0}^{\infty} (-1)^n \frac{z^{2n+1}}{(2n+1)!} \qquad (|z| < \infty)$$

to see that when $0 < |z| < \infty$,

$$z^2 \sin\left(\frac{1}{z^2}\right) = \sum_{n=0}^{\infty} \frac{(-1)^n}{(2n+1)!} \cdot \frac{1}{z^{4n}} = 1 + \sum_{n=1}^{\infty} \frac{(-1)^n}{(2n+1)!} \cdot \frac{1}{z^{4n}}.$$

3. Suppose that $1 < |z| < \infty$ and recall the Maclaurin series representation

$$\frac{1}{1-z} = \sum_{n=0}^{\infty} z^n \qquad (|z| < 1).$$

This enables us to write

$$\frac{1}{1+z} = \frac{1}{z} \cdot \frac{1}{1+\frac{1}{z}} = \frac{1}{z} \sum_{n=0}^{\infty} \left(-\frac{1}{z}\right)^n = \sum_{n=0}^{\infty} \frac{(-1)^n}{z^{n+1}} \qquad (1 < |z| < \infty).$$

Replacing n by $n-1$ in this last series and then noting that

$$(-1)^{n-1} = (-1)^{n-1}(-1)^2 = (-1)^{n+1},$$

we arrive at the desired expansion:

$$\frac{1}{1+z} = \sum_{n=1}^{\infty} \frac{(-1)^{n+1}}{z^n} \qquad (1 < |z| < \infty).$$

4. The singularities of the function $f(z) = \dfrac{1}{z^2(1-z)}$ are at the points $z = 0$ and $z = 1$. Hence there are Laurent series in powers of z for the domains $0 < |z| < 1$ and $1 < |z| < \infty$ (see the figure below).

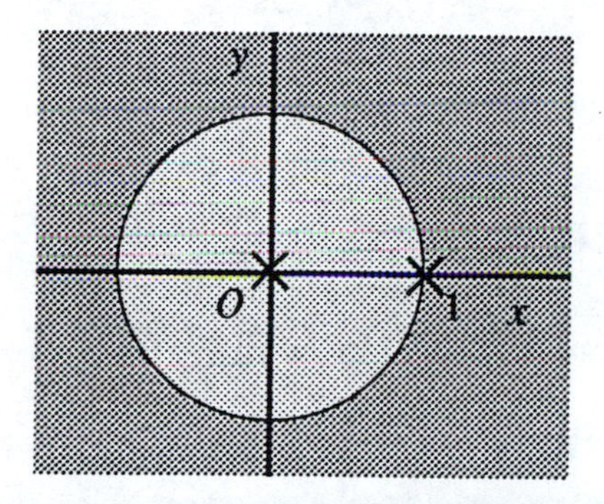

To find the series when $0 < |z| < 1$, recall that $\dfrac{1}{1-z} = \sum_{n=0}^{\infty} z^n$ $(|z| < 1)$ and write

$$f(z) = \frac{1}{z^2} \cdot \frac{1}{1-z} = \frac{1}{z^2} \sum_{n=0}^{\infty} z^n = \sum_{n=0}^{\infty} z^{n-2} = \frac{1}{z^2} + \frac{1}{z} + \sum_{n=2}^{\infty} z^{n-2} = \sum_{n=0}^{\infty} z^n + \frac{1}{z} + \frac{1}{z^2}.$$

As for the domain $1 < |z| < \infty$, note that $|1/z| < 1$ and write

$$f(z) = -\frac{1}{z^3} \cdot \frac{1}{1-(1/z)} = -\frac{1}{z^3} \sum_{n=0}^{\infty} \left(\frac{1}{z}\right)^n = -\sum_{n=0}^{\infty} \frac{1}{z^{n+3}} = -\sum_{n=3}^{\infty} \frac{1}{z^n}.$$

5. *(a)* The Maclaurin series for the function $\dfrac{z+1}{z-1}$ is valid when $|z| < 1$. To find it, we recall the Maclaurin series representation

$$\frac{1}{1-z} = \sum_{n=0}^{\infty} z^n \qquad (|z| < 1)$$

for $\dfrac{1}{1-z}$ and write

$$\frac{z+1}{z-1} = -(z+1)\frac{1}{1-z} = (-z-1)\sum_{n=0}^{\infty} z^n = -\sum_{n=0}^{\infty} z^{n+1} - \sum_{n=0}^{\infty} z^n$$
$$= -\sum_{n=1}^{\infty} z^n - \sum_{n=0}^{\infty} z^n = -1 - 2\sum_{n=1}^{\infty} z^n \qquad (|z| < 1).$$

(b) To find the Laurent series for the same function when $1<|z|<\infty$, we recall the Maclaurin series for $\dfrac{1}{1-z}$ that was used in part *(a)*. Since $\left|\dfrac{1}{z}\right|<1$ here, we may write

$$\frac{z+1}{z-1}=\frac{1+\dfrac{1}{z}}{1-\dfrac{1}{z}}=\left(1+\frac{1}{z}\right)\frac{1}{1-\dfrac{1}{z}}=\left(1+\frac{1}{z}\right)\sum_{n=0}^{\infty}\left(\frac{1}{z}\right)^n=\sum_{n=0}^{\infty}\frac{1}{z^n}+\sum_{n=0}^{\infty}\frac{1}{z^{n+1}}$$

$$=\sum_{n=0}^{\infty}\frac{1}{z^n}+\sum_{n=1}^{\infty}\frac{1}{z^n}=1+2\sum_{n=1}^{\infty}\frac{1}{z^n} \qquad (1<|z|<\infty).$$

7. The function $f(z)=\dfrac{1}{z(1+z^2)}$ has isolated singularities at $z=0$ and $z=\pm i$, as indicated in the figure below. Hence there is a Laurent series representation for the domain $0<|z|<1$ and also one for the domain $1<|z|<\infty$, which is exterior to the circle $|z|=1$.

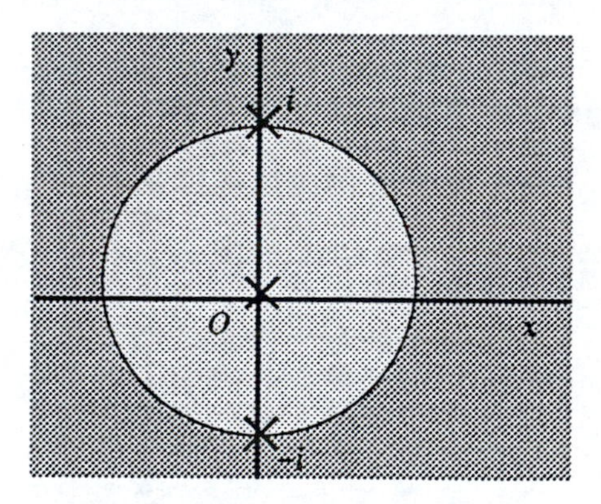

To find each of these Laurent series, we recall the Maclaurin series representation

$$\frac{1}{1-z}=\sum_{n=0}^{\infty}z^n \qquad (|z|<1).$$

For the domain $0<|z|<1$, we have

$$f(z)=\frac{1}{z}\cdot\frac{1}{1+z^2}=\frac{1}{z}\sum_{n=0}^{\infty}(-z^2)^n=\sum_{n=0}^{\infty}(-1)^n z^{2n-1}=\frac{1}{z}+\sum_{n=1}^{\infty}(-1)^n z^{2n-1}=\sum_{n=0}^{\infty}(-1)^{n+1}z^{2n+1}+\frac{1}{z}.$$

On the other hand, when $1<|z|<\infty$,

$$f(z)=\frac{1}{z^3}\cdot\frac{1}{1+\dfrac{1}{z^2}}=\frac{1}{z^3}\sum_{n=0}^{\infty}\left(-\frac{1}{z^2}\right)^n=\sum_{n=0}^{\infty}\frac{(-1)^n}{z^{2n+3}}=\sum_{n=1}^{\infty}\frac{(-1)^{n+1}}{z^{2n+1}}.$$

In this second expansion, we have used the fact that $(-1)^{n-1}=(-1)^{n-1}(-1)^2=(-1)^{n+1}$.

8. *(a)* Let a denote a real number, where $-1 < a < 1$. Recalling that

$$\frac{1}{1-z} = \sum_{n=0}^{\infty} z^n \qquad (|z| < 1)$$

enables us to write

$$\frac{a}{z-a} = \frac{a}{z} \cdot \frac{1}{1-(a/z)} = \sum_{n=0}^{\infty} \frac{a^{n+1}}{z^{n+1}},$$

or

$$\frac{a}{z-a} = \sum_{n=1}^{\infty} \frac{a^n}{z^n} \qquad (|a| < |z| < \infty).$$

(b) Putting $z = e^{i\theta}$ on each side of the final result in part *(a)*, we have

$$\frac{a}{e^{i\theta} - a} = \sum_{n=1}^{\infty} a^n e^{-in\theta}.$$

But

$$\frac{a}{e^{i\theta} - a} = \frac{a}{(\cos\theta - a) + i\sin\theta} \cdot \frac{(\cos\theta - a) - i\sin\theta}{(\cos\theta - a) - i\sin\theta} = \frac{a\cos\theta - a^2 - ia\sin\theta}{1 - 2a\cos\theta + a^2}$$

and

$$\sum_{n=1}^{\infty} a^n e^{-in\theta} = \sum_{n=1}^{\infty} a^n \cos n\theta - i\sum_{n=1}^{\infty} a^n \sin n\theta.$$

Consequently,

$$\sum_{n=1}^{\infty} a^n \cos n\theta = \frac{a\cos\theta - a^2}{1 - 2a\cos\theta + a^2} \quad \text{and} \quad \sum_{n=1}^{\infty} a^n \sin n\theta = \frac{a\sin\theta}{1 - 2a\cos\theta + a^2}$$

when $-1 < a < 1$.

10. *(a)* Let z be any fixed complex number and C the unit circle $w = e^{i\phi}$ $(-\pi \le \phi \le \pi)$ in the w plane. The function

$$f(w) = \exp\left[\frac{z}{2}\left(w - \frac{1}{w}\right)\right]$$

has the one singularity $w = 0$ in the w plane. That singularity is, of course, interior to C, as shown in the figure below.

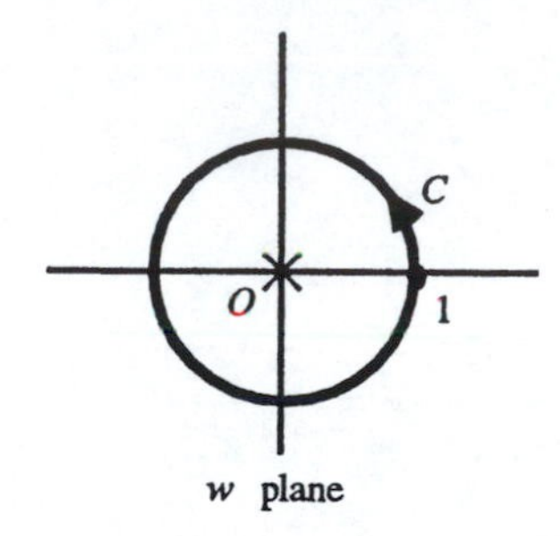

Now the function $f(w)$ has a Laurent series representation in the domain $0<|w|<\infty$. According to expression (5), Sec. 55, then,

$$\exp\left[\frac{z}{2}\left(w-\frac{1}{w}\right)\right]=\sum_{n=-\infty}^{\infty}J_n(z)w^n \qquad (0<|w|<\infty),$$

where the coefficients $J_n(z)$ are

$$J_n(z)=\frac{1}{2\pi i}\int_C \frac{\exp\left[\frac{z}{2}\left(w-\frac{1}{w}\right)\right]}{w^{n+1}}dw \qquad (n=0,\pm1,\pm2,\ldots).$$

Using the parametric representation $w=e^{i\phi}$ $(-\pi\le\phi\le\pi)$ for C, let us rewrite this expression for $J_n(z)$ as follows:

$$J_n(z)=\frac{1}{2\pi i}\int_{-\pi}^{\pi}\frac{\exp\left[\frac{z}{2}\left(e^{i\phi}-e^{-i\phi}\right)\right]}{e^{i(n+1)\phi}}ie^{i\phi}d\phi=\frac{1}{2\pi i}\int_{-\pi}^{\pi}\exp[iz\sin\phi]e^{-in\phi}d\phi.$$

That is,

$$J_n(z)=\frac{1}{2\pi}\int_{-\pi}^{\pi}\exp[-i(n\phi-z\sin\phi)]d\phi \qquad (n=0,\pm1,\pm2,\ldots).$$

(b) The last expression for $J_n(z)$in part *(a)* can be written as

$$J_n(z)=\frac{1}{2\pi}\int_{-\pi}^{\pi}[\cos(n\phi-z\sin\phi)-i\sin(n\phi-z\sin\phi)]d\phi$$

$$=\frac{1}{2\pi}\int_{-\pi}^{\pi}\cos(n\phi-z\sin\phi)d\phi-\frac{i}{2\pi}\int_{-\pi}^{\pi}\sin(n\phi-z\sin\phi)d\phi$$

$$=\frac{1}{2\pi}2\int_{0}^{\pi}\cos(n\phi-z\sin\phi)d\phi-\frac{i}{2\pi}0 \qquad (n=0,\pm1,\pm2,\ldots).$$

That is,

$$J_n(z) = \frac{1}{\pi}\int_0^{\pi} \cos(n\phi - z\sin\phi)\,d\phi \qquad (n = 0, \pm 1, \pm 2, \ldots).$$

11. *(a)* The function $f(z)$ is analytic in some annular domain centered at the origin; and the unit circle $C: z = e^{i\phi}\ (-\pi \le \phi \le \pi)$ is contained in that domain, as shown below.

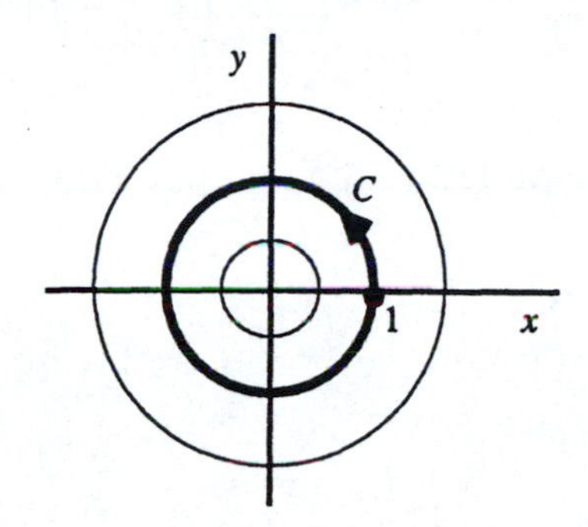

For each point z in the annular domain, there is a Laurent series representation

$$f(z) = \sum_{n=0}^{\infty} a_n z^n + \sum_{n=1}^{\infty} \frac{b_n}{z^n},$$

where

$$a_n = \frac{1}{2\pi i}\int_C \frac{f(z)\,dz}{z^{n+1}} = \frac{1}{2\pi i}\int_{-\pi}^{\pi} \frac{f(e^{i\phi})}{e^{i\phi(n+1)}} ie^{i\phi}\,d\phi = \frac{1}{2\pi}\int_{-\pi}^{\pi} f(e^{i\phi})e^{-in\phi}\,d\phi \qquad (n = 0, 1, 2, \ldots)$$

and

$$b_n = \frac{1}{2\pi i}\int_C \frac{f(z)\,dz}{z^{-n+1}} = \frac{1}{2\pi i}\int_{-\pi}^{\pi} \frac{f(e^{i\phi})}{e^{i\phi(-n+1)}} ie^{i\phi}\,d\phi = \frac{1}{2\pi}\int_{-\pi}^{\pi} f(e^{i\phi})e^{in\phi}\,d\phi \qquad (n = 1, 2, \ldots).$$

Substituting these values of a_n and b_n into the series, we then have

$$f(z) = \sum_{n=0}^{\infty} \frac{1}{2\pi}\int_{-\pi}^{\pi} f(e^{i\phi})e^{-in\phi}\,d\phi\; z^n + \sum_{n=1}^{\infty} \frac{1}{2\pi}\int_{-\pi}^{\pi} f(e^{i\phi})e^{in\phi}\,d\phi\; \frac{1}{z^n},$$

or

$$f(z) = \frac{1}{2\pi}\int_{-\pi}^{\pi} f(e^{i\phi})\,d\phi + \frac{1}{2\pi}\sum_{n=1}^{\infty}\int_{-\pi}^{\pi} f(e^{i\phi})\left[\left(\frac{z}{e^{i\phi}}\right)^n + \left(\frac{e^{i\phi}}{z}\right)^n\right]d\phi.$$

(b) Put $z = e^{i\theta}$ in the final result in part *(a)* to get

$$f(e^{i\theta}) = \frac{1}{2\pi}\int_{-\pi}^{\pi} f(e^{i\phi})d\phi + \frac{1}{2\pi}\sum_{n=1}^{\infty}\int_{-\pi}^{\pi} f(e^{i\phi})[e^{in(\theta-\phi)} + e^{-in(\theta-\phi)}]d\phi,$$

or

$$f(e^{i\theta}) = \frac{1}{2\pi}\int_{-\pi}^{\pi} f(e^{i\phi})d\phi + \frac{1}{\pi}\sum_{n=1}^{\infty}\int_{-\pi}^{\pi} f(e^{i\phi})\cos[n(\theta-\phi)]d\phi.$$

If $u(\theta) = \operatorname{Re} f(e^{i\theta})$, then, equating the real parts on each side of this last equation yields

$$u(\theta) = \frac{1}{2\pi}\int_{-\pi}^{\pi} u(\phi)d\phi + \frac{1}{\pi}\sum_{n=1}^{\infty}\int_{-\pi}^{\pi} u(\phi)\cos[n(\theta-\phi)]d\phi.$$

SECTION 60

1. Differentiating each side of the representation

$$\frac{1}{1-z} = \sum_{n=0}^{\infty} z^n \qquad (|z| < 1),$$

we find that

$$\frac{1}{(1-z)^2} = \frac{d}{dz}\sum_{n=0}^{\infty} z^n = \sum_{n=0}^{\infty}\frac{d}{dz}z^n = \sum_{n=1}^{\infty} nz^{n-1} = \sum_{n=0}^{\infty}(n+1)z^n \qquad (|z| < 1).$$

Another differentiation gives

$$\frac{2}{(1-z)^3} = \frac{d}{dz}\sum_{n=0}^{\infty}(n+1)z^n = \sum_{n=0}^{\infty}(n+1)\frac{d}{dz}z^n = \sum_{n=1}^{\infty} n(n+1)z^{n-1} = \sum_{n=0}^{\infty}(n+1)(n+2)z^n \qquad (|z| < 1).$$

2. Replace z by $1/(1-z)$ on each side of the Maclaurin series representation (Exercise 1)

$$\frac{1}{(1-z)^2} = \sum_{n=0}^{\infty}(n+1)z^n \qquad (|z| < 1),$$

as well as in its condition of validity. This yields the Laurent series representation

$$\frac{1}{z^2} = \sum_{n=2}^{\infty}\frac{(-1)^n(n-1)}{(z-1)^n} \qquad (1 < |z-1| < \infty).$$

3. Since the function $f(z)=1/z$ has a singular point at $z=0$, its Taylor series about $z_0=2$ is valid in the open disk $|z-2|<2$, as indicated in the figure below.

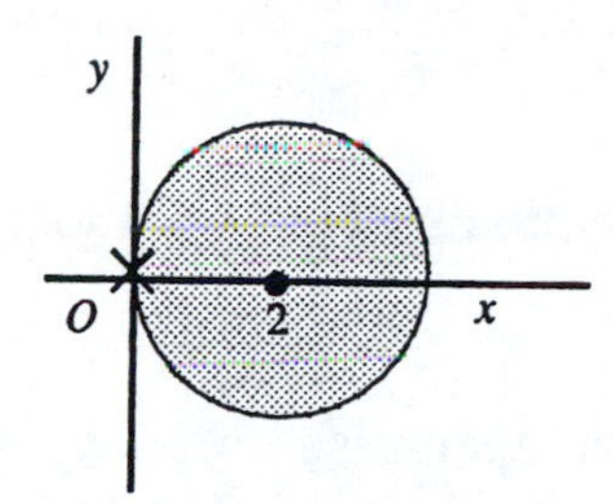

To find that series, write

$$\frac{1}{z}=\frac{1}{2+(z-2)}=\frac{1}{2}\cdot\frac{1}{1+(z-2)/2}$$

to see that it can be obtained by replacing z by $-(z-2)/2$ in the known expansion

$$\frac{1}{1-z}=\sum_{n=0}^{\infty}z^n \qquad (|z|<1).$$

Specifically,

$$\frac{1}{z}=\frac{1}{2}\sum_{n=0}^{\infty}\left[-\frac{(z-2)}{2}\right]^n \qquad (|z-2|<2),$$

or

$$\frac{1}{z}=\sum_{n=0}^{\infty}\frac{(-1)^n}{2^{n+1}}(z-2)^n \qquad (|z-2|<2).$$

Differentiating this series term by term, we have

$$-\frac{1}{z^2}=\sum_{n=1}^{\infty}\frac{(-1)^n}{2^{n+1}}n(z-2)^{n-1}=\sum_{n=0}^{\infty}\frac{(-1)^{n+1}}{2^{n+2}}(n+1)(z-2)^n \qquad (|z-2|<2).$$

Thus

$$\frac{1}{z^2}=\frac{1}{4}\sum_{n=0}^{\infty}(-1)^n(n+1)\left(\frac{z-2}{2}\right)^n \qquad (|z-2|<2).$$

4. Consider the function defined by the equations

$$f(z)=\begin{cases}\dfrac{e^z-1}{z} & \text{when } z\neq 0,\\ 1 & \text{when } z=0.\end{cases}$$

When $z \neq 0$, $f(z)$ has the power series representation

$$f(z) = \frac{1}{z}\left[\left(1 + \frac{z}{1!} + \frac{z^2}{2!} + \frac{z^3}{3!} + \cdots\right) - 1\right] = 1 + \frac{z}{2!} + \frac{z^2}{3!} + \cdots.$$

Since this representation clearly holds when $z = 0$ too, it is actually valid for all z. Hence f is entire.

6. Let C be a contour lying in the open disk $|w-1| < 1$ in the w plane that extends from the point $w = 1$ to a point $w = z$, as shown in the figure below.

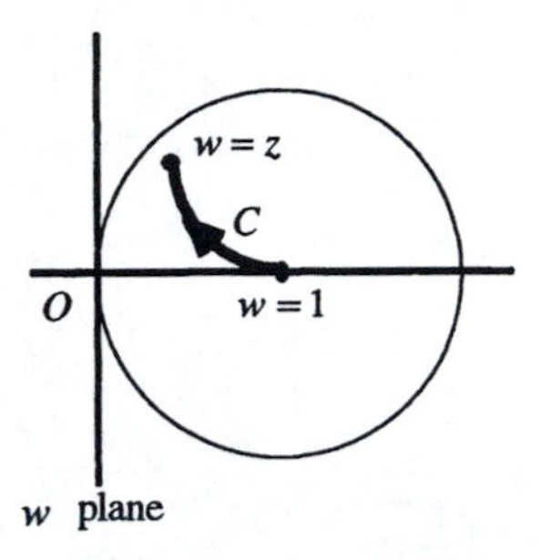

According to Theorem 1 in Sec. 59, we can integrate the Taylor series representation

$$\frac{1}{w} = \sum_{n=0}^{\infty} (-1)^n (w-1)^n \qquad (|w-1| < 1)$$

term by term along the contour C. Thus

$$\int_C \frac{dw}{w} = \int_C \sum_{n=0}^{\infty} (-1)^n (w-1)^n \, dw = \sum_{n=0}^{\infty} (-1)^n \int_C (w-1)^n \, dw.$$

But

$$\int_C \frac{dw}{w} = \int_1^z \frac{dw}{w} = [\operatorname{Log} w]_1^z = \operatorname{Log} z - \operatorname{Log} 1 = \operatorname{Log} z$$

and

$$\int_C (w-1)^n = \int_1^z (w-1)^n \, dw = \left[\frac{(w-1)^{n+1}}{n+1}\right]_1^z = \frac{(z-1)^{n+1}}{n+1}.$$

Hence

$$\operatorname{Log} z = \sum_{n=0}^{\infty} \frac{(-1)^n}{n+1} (z-1)^{n+1} = \sum_{n=1}^{\infty} \frac{(-1)^{n-1}}{n} (z-1)^n \qquad (|z-1| < 1);$$

and, since $(-1)^{n-1} = (-1)^{n-1}(-1)^2 = (-1)^{n+1}$, this result becomes

$$\operatorname{Log} z = \sum_{n=1}^{\infty} \frac{(-1)^{n+1}}{n} (z-1)^n \qquad (|z-1| < 1).$$

SECTION 61

1. The singularities of the function $f(z)=\dfrac{e^z}{z(z^2+1)}$ are at $z=0,\pm i$. The problem here is to find the Laurent series for f that is valid in the punctured disk $0<|z|<1$, shown below.

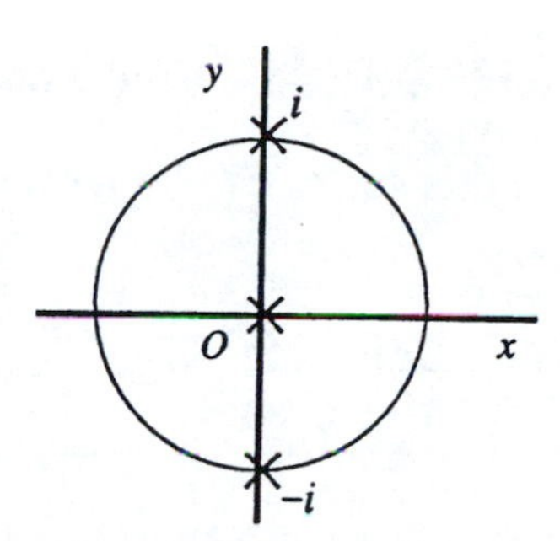

We begin by recalling the Maclaurin series representations

$$e^z=1+\frac{z}{1!}+\frac{z^2}{2!}+\frac{z^3}{3!}+\cdots \qquad (|z|<\infty)$$

and

$$\frac{1}{1-z}=1+z+z^2+z^3+\cdots \qquad (|z|<1),$$

which enable us to write

$$e^z=1+z+\frac{1}{2}z^2+\frac{1}{6}z^3+\cdots \qquad (|z|<\infty)$$

and

$$\frac{1}{z^2+1}=1-z^2+z^4-z^6+\cdots \qquad (|z|<1).$$

Multiplying these last two series term by term, we have the Maclaurin series representation

$$\begin{aligned}\frac{e^z}{z^2+1}&=1+z+\frac{1}{2}z^2+\frac{1}{6}z^3+\cdots\\ &\qquad\qquad -z^2\quad -z^3-\cdots\\ &\qquad\qquad\qquad\qquad\quad z^4+\cdots\\ &\qquad\qquad\qquad\qquad\qquad\quad \vdots\\ &=1+z-\frac{1}{2}z^2-\frac{5}{6}z^3+\cdots,\end{aligned}$$

which is valid when $|z|<1$. The desired Laurent series is then obtained by multiplying each side of the above representation by $\dfrac{1}{z}$:

$$\frac{e^z}{z(z^2+1)}=\frac{1}{z}+1-\frac{1}{2}z-\frac{5}{6}z^2+\cdots \qquad (0<|z|<1).$$

4. We know the Laurent series representation

$$\frac{1}{z^2 \sinh z} = \frac{1}{z^3} - \frac{1}{6} \cdot \frac{1}{z} + \frac{7}{360} z + \cdots \qquad (0 < |z| < \pi)$$

from Example 2, Sec. 61. Expression (3), Sec. 55, for the coefficients b_n in a Laurent series tells us that the coefficient b_1 of $\frac{1}{z}$ in this series can be written

$$b_1 = \frac{1}{2\pi i} \int_C \frac{dz}{z^2 \sinh z},$$

where C is the circle $|z| = 1$, taken counterclockwise. Since $b_1 = -\frac{1}{6}$, then,

$$\int_C \frac{dz}{z^2 \sinh z} = 2\pi i\left(-\frac{1}{6}\right) = -\frac{\pi i}{3}.$$

6. The problem here is to use mathematical induction to verify the differentiation formula

$$[f(z)g(z)]^{(n)} = \sum_{k=0}^{n} \binom{n}{k} f^{(k)}(z) g^{(n-k)}(z) \qquad (n = 1, 2, \ldots).$$

The formula is clearly true when $n = 1$ since in that case it becomes

$$[f(z)g(z)]' = f(z)g'(z) + f'(z)g(z).$$

We now assume that the formula is true when $n = m$ and show how, as a consequence, it is true when $n = m + 1$. We start by writing

$$[f(z)g(z)]^{(m+1)} = \{[f(z)g(z)]'\}^{(m)} = [f(z)g'(z) + f'(z)g(z)]^{(m)}$$

$$= [f(z)g'(z)]^{(m)} + [f'(z)g(z)]^{(m)}$$

$$= \sum_{k=0}^{m} \binom{m}{k} f^{(k)}(z) g^{(m-k+1)}(z) + \sum_{k=0}^{m} \binom{m}{k} f^{(k+1)}(z) g^{(m-k)}(z)$$

$$= \sum_{k=0}^{m} \binom{m}{k} f^{(k)}(z) g^{(m-k+1)}(z) + \sum_{k=1}^{m+1} \binom{m}{k-1} f^{(k)}(z) g^{(m-k+1)}(z)$$

$$= f(z)g^{(m+1)}(z) + \sum_{k=1}^{m} \left[\binom{m}{k} + \binom{m}{k-1}\right] f^{(k)}(z) g^{(m+1-k)}(z) + f^{(m+1)}(z)g(z).$$

But

$$\binom{m}{k}+\binom{m}{k-1}=\frac{m!}{k!(m-k)!}+\frac{m!}{(k-1)!(m-k+1)!}=\frac{(m+1)!}{k!(m+1-k)!}=\binom{m+1}{k};$$

and so

$$[f(z)g(z)]^{(m+1)}=f(z)g^{(m+1)}(z)+\sum_{k=1}^{m}\binom{m+1}{k}f^{(k)}(z)g^{(m+1-k)}(z)+f^{(m+1)}(z)g(z),$$

or

$$[f(z)g(z)]^{(m+1)}=\sum_{k=0}^{m+1}\binom{m+1}{k}f^{(k)}(z)g^{(m+1-k)}(z).$$

The desired verification is now complete.

7. We are given that $f(z)$ is an entire function represented by a series of the form

$$f(z)=z+a_2z^2+a_3z^3+\cdots \qquad (|z|<\infty).$$

(a) Write $g(z)=f[f(z)]$ and observe that

$$f[f(z)]=g(0)+\frac{g'(0)}{1!}z+\frac{g''(0)}{2!}z^2+\frac{g'''(0)}{3!}z^3+\cdots \qquad (|z|<\infty).$$

It is straightforward to show that

$$g'(z)=f'[f(z)]f'(z),$$

$$g''(z)=f''[f(z)][f'(z)]^2+f'[f(z)]f''(z),$$

and

$$g'''(z)=f'''[f(z)][f'(z)]^3+2f'(z)f''(z)f''[f(z)]+f''[f(z)]f'(z)f''(z)+f'[f(z)]f'''(z).$$

Thus

$$g(0)=0,\quad g'(0)=1,\quad g''(0)=4a_2,\quad \text{and}\quad g'''(0)=12(a_2^2+a_3),$$

and so

$$f[f(z)]=z+2a_2z^2+2(a_2^2+a_3)z^3+\cdots \qquad (|z|<\infty).$$

(*b*) Proceeding formally, we have

$$f[f(z)] = f(z) + a_2[f(z)]^2 + a_3[f(z)]^3 + \cdots$$

$$= (z + a_2z^2 + a_3z^3 + \cdots) + a_2(z + a_2z^2 + a_3z^3 + \cdots)^2 + a_3(z + a_2z^2 + a_3z^3 + \cdots)^3 + \cdots$$

$$= (z + a_2z^2 + a_3z^3 + \cdots) + (a_2z^2 + 2a_2^2z^3 + \cdots) + (a_3z^3 + \cdots)$$

$$= z + 2a_2z^2 + 2(a_2^2 + a_3)z^3 + \cdots.$$

(*c*) Since

$$\sin z = z - \frac{z^3}{3!} + \cdots = z + 0z^2 + \left(-\frac{1}{6}\right)z^3 + \cdots \qquad (|z| < \infty),$$

the result in part (*a*), with $a_2 = 0$ and $a_3 = -\frac{1}{6}$, tells us that

$$\sin(\sin z) = z - \frac{1}{3}z^3 + \cdots \qquad (|z| < \infty).$$

8. We need to find the first four nonzero coefficients in the Maclaurin series representation

$$\frac{1}{\cosh z} = \sum_{n=0}^{\infty} \frac{E_n}{n!} z^n \qquad \left(|z| < \frac{\pi}{2}\right).$$

This representation is valid in the stated disk since the zeros of $\cosh z$ are the numbers $z = \left(\frac{\pi}{2} + n\pi\right)i$ $(n = 0, \pm1, \pm2, \ldots)$, the ones nearest to the origin being $z = \pm\frac{\pi}{2}i$. The series contains only even powers of z since $\cosh z$ is an even function; that is, $E_{2n+1} = 0$ $(n = 0, 1, 2, \ldots)$. To find the series, we divide the series

$$\cosh z = 1 + \frac{z^2}{2!} + \frac{z^4}{4!} + \frac{z^6}{6!} + \cdots = 1 + \frac{1}{2}z^2 + \frac{1}{24}z^4 + \frac{1}{720}z^6 + \cdots \qquad (|z| < \infty)$$

into 1. The result is

$$\frac{1}{\cosh z} = 1 - \frac{1}{2}z^2 + \frac{5}{24}z^4 - \frac{61}{720}z^6 + \cdots \qquad \left(|z| < \frac{\pi}{2}\right),$$

or

$$\frac{1}{\cosh z} = 1 - \frac{1}{2!}z^2 + \frac{5}{4!}z^4 - \frac{61}{6!}z^6 + \cdots \qquad \left(|z| < \frac{\pi}{2}\right).$$

Since

$$\frac{1}{\cosh z} = E_0 + \frac{E_2}{2!}z^2 + \frac{E_4}{4!}z^4 + \frac{E_6}{6!}z^6 + \cdots \qquad \left(|z| < \frac{\pi}{2}\right),$$

this tells us that

$$E_0 = 1, \quad E_2 = -1, \quad E_4 = 5, \quad \text{and} \quad E_6 = -61.$$

Chapter 6

SECTION 64

1. *(a)* Let us write

$$\frac{1}{z+z^2}=\frac{1}{z}\cdot\frac{1}{1+z}=\frac{1}{z}\left(1-z+z^2-z^3+\cdots\right)=\frac{1}{z}-1+z-z^2+\cdots \qquad (0<|z|<1).$$

The residue at $z=0$, which is the coefficient of $\frac{1}{z}$, is clearly 1.

(b) We may use the expansion

$$\cos z=1-\frac{z^2}{2!}+\frac{z^4}{4!}-\frac{z^6}{6!}+\cdots \qquad (|z|<\infty)$$

to write

$$z\cos\left(\frac{1}{z}\right)=z\left(1-\frac{1}{2!}\cdot\frac{1}{z^2}+\frac{1}{4!}\cdot\frac{1}{z^4}-\frac{1}{6!}\cdot\frac{1}{z^6}+\cdots\right)=z-\frac{1}{2!}\cdot\frac{1}{z}+\frac{1}{4!}\cdot\frac{1}{z^3}-\frac{1}{6!}\cdot\frac{1}{z^5}+\cdots$$

$$(0<|z|<\infty).$$

The residue at $z=0$, or coefficient of $\frac{1}{z}$, is now seen to be $-\frac{1}{2}$.

(c) Observe that

$$\frac{z-\sin z}{z}=\frac{1}{z}(z-\sin z)=\frac{1}{z}\left[z-\left(z-\frac{z^3}{3!}+\frac{z^5}{5!}-\cdots\right)\right]=\frac{z^2}{3!}-\frac{z^4}{5!}+\cdots \qquad (0<|z|<\infty).$$

Since the coefficient of $\frac{1}{z}$ in this Laurent series is 0, the residue at $z=0$ is 0.

(d) Write

$$\frac{\cot z}{z^4}=\frac{1}{z^4}\cdot\frac{\cos z}{\sin z}$$

and recall that

$$\cos z=1-\frac{z^2}{2!}+\frac{z^4}{4!}-\cdots=1-\frac{z^2}{2}+\frac{z^4}{24}-\cdots \qquad (|z|<\infty)$$

and

$$\sin z=z-\frac{z^3}{3!}+\frac{z^5}{5!}-\cdots=z-\frac{z^3}{6}+\frac{z^5}{120}-\cdots \qquad (|z|<\infty).$$

Dividing the series for $\sin z$ into the one for $\cos z$, we find that

$$\frac{\cos z}{\sin z} = \frac{1}{z} - \frac{z}{3} - \frac{z^3}{45} + \cdots \qquad (0 < |z| < \pi).$$

Thus

$$\frac{\cot z}{z^4} = \frac{1}{z^4}\left(\frac{1}{z} - \frac{z}{3} - \frac{z^3}{45} + \cdots\right) = \frac{1}{z^5} - \frac{1}{3}\cdot\frac{1}{z^3} - \frac{1}{45}\cdot\frac{1}{z} + \cdots \qquad (0 < |z| < \pi).$$

Note that the condition of validity for this series is due to the fact that $\sin z = 0$ when $z = n\pi$ $(n = 0, \pm 1, \pm 2, \ldots)$. It is now evident that $\frac{\cot z}{z^4}$ has residue $-\frac{1}{45}$. at $z = 0$.

(e) Recall that

$$\sinh z = z + \frac{z^3}{3!} + \frac{z^5}{5!} + \cdots \qquad (|z| < \infty)$$

and

$$\frac{1}{1-z} = 1 + z + z^2 + \cdots \qquad (|z| < \infty).$$

There is a Laurent series for the function

$$\frac{\sinh z}{z^4(1-z^2)} = \frac{1}{z^4}\cdot(\sinh z)\left(\frac{1}{1-z^2}\right)$$

that is valid for $0 < |z| < 1$. To find it, we first multiply the Maclaurin series for $\sinh z$ and $\frac{1}{1-z^2}$:

$$\begin{aligned}(\sinh z)\left(\frac{1}{1-z^2}\right) &= \left(z + \frac{1}{6}z^3 + \frac{1}{120}z^5 + \cdots\right)\left(1 + z^2 + z^4 + \cdots\right) \\ &= z + \frac{1}{6}z^3 + \frac{1}{120}z^5 + \cdots \\ &\qquad\quad z^3 + \frac{1}{6}z^5 + \cdots \\ &\qquad\qquad\qquad z^5 + \cdots \\ &= z + \frac{7}{6}z^3 + \cdots \qquad (0 < |z| < 1).\end{aligned}$$

We then see that

$$\frac{\sinh z}{z^4(1-z^2)} = \frac{1}{z^3} + \frac{7}{6}\cdot\frac{1}{z} + \cdots \qquad (0 < |z| < 1).$$

This shows that the residue of $\dfrac{\sinh z}{z^4(1-z^2)}$ at $z = 0$ is $\dfrac{7}{6}$.

2. In each part, C denotes the positively oriented circle $|z| = 3$.

(a) To evaluate $\int_C \frac{\exp(-z)}{z^2}dz$, we need the residue of the integrand at $z = 0$. From the Laurent series

$$\frac{\exp(-z)}{z^2} = \frac{1}{z^2}\left(1 - \frac{z}{1!} + \frac{z^2}{2!} - \frac{z^3}{3!} + \cdots\right) = \frac{1}{z^2} - \frac{1}{1!}\cdot\frac{1}{z} + \frac{1}{2!} - \frac{z}{3!} + \cdots \qquad (0 < |z| < \infty),$$

we see that the required residue is -1. Thus

$$\int_C \frac{\exp(-z)}{z^2}dz = 2\pi i(-1) = -2\pi i.$$

(c) Likewise, to evaluate the integral $\int_C z^2 \exp\left(\frac{1}{z}\right)dz$, we must find the residue of the integrand at $z = 0$. The Laurent series

$$z^2 \exp\left(\frac{1}{z}\right) = z^2\left(1 + \frac{1}{1!}\cdot\frac{1}{z} + \frac{1}{2!}\cdot\frac{1}{z^2} + \frac{1}{3!}\cdot\frac{1}{z^3} + \frac{1}{4!}\cdot\frac{1}{z^4} + \cdots\right)$$

$$= z^2 + \frac{z}{1!} + \frac{1}{2!} + \frac{1}{3!}\cdot\frac{1}{z} + \frac{1}{4!}\cdot\frac{1}{z^2} + \cdots,$$

which is valid for $0 < |z| < \infty$, tells us that the needed residue is $\dfrac{1}{6}$. Hence

$$\int_C z^2 \exp\left(\frac{1}{z}\right)dz = 2\pi i\left(\frac{1}{6}\right) = \frac{\pi i}{3}.$$

(d) As for the integral $\int_C \frac{z+1}{z^2-2z}dz$, we need the two residues of

$$\frac{z+1}{z^2-2z} = \frac{z+1}{z(z-2)},$$

one at $z=0$ and one at $z=2$. The residue at $z=0$ can be found by writing

$$\frac{z+1}{z(z-2)} = \left(\frac{z+1}{z}\right)\left(\frac{1}{z-2}\right) = \left(-\frac{1}{2}\right)\left(1+\frac{1}{z}\right)\cdot\frac{1}{1-(z/2)}$$

$$= \left(-\frac{1}{2}-\frac{1}{2}\cdot\frac{1}{z}\right)\left(1+\frac{z}{2}+\frac{z^2}{2^2}+\cdots\right),$$

which is valid when $0<|z|<2$, and observing that the coefficient of $\frac{1}{z}$ in this last product is $-\frac{1}{2}$. To obtain the residue at $z=2$, we write

$$\frac{z+1}{z(z-2)} = \frac{(z-2)+3}{z-2}\cdot\frac{1}{2+(z-2)} = \frac{1}{2}\left(1+\frac{3}{z-2}\right)\cdot\frac{1}{1+(z-2)/2}$$

$$= \frac{1}{2}\left(1+\frac{3}{z-2}\right)\left[1-\frac{z-2}{2}+\frac{(z-2)^2}{2^2}-\cdots\right],$$

which is valid when $0<|z-2|<2$, and note that the coefficient of $\frac{1}{z-2}$ in this product is $\frac{3}{2}$. Finally, then, by the residue theorem,

$$\int_C \frac{z+1}{z^2-2z}dz = 2\pi i\left(-\frac{1}{2}+\frac{3}{2}\right) = 2\pi i.$$

3. In each part of this problem, C is the positively oriented circle $|z|=2$.

(a) If $f(z)=\frac{z^5}{1-z^3}$, then

$$\frac{1}{z^2}f\left(\frac{1}{z}\right) = \frac{1}{z^7-z^4} = -\frac{1}{z^4}\cdot\frac{1}{1-z^3} = -\frac{1}{z^4}(1+z^3+z^6+\cdots) = -\frac{1}{z^4}-\frac{1}{z}-z^2-\cdots$$

when $0<|z|<1$. This tells us that

$$\int_C f(z)\,dz = 2\pi i \operatorname*{Res}_{z=0} \frac{1}{z^2}f\left(\frac{1}{z}\right) = 2\pi i(-1) = -2\pi i.$$

(b) When $f(z)=\dfrac{1}{1+z^2}$, we have

$$\frac{1}{z^2}f\left(\frac{1}{z}\right)=\frac{1}{1+z^2}=\frac{1}{1-(-z^2)}=1-z^2+z^4-\cdots \qquad (0<|z|<1).$$

Thus

$$\int_C f(z)\,dz=2\pi i \operatorname*{Res}_{z=0}\frac{1}{z^2}f\left(\frac{1}{z}\right)=2\pi i(0)=0.$$

(c) If $f(z)=\dfrac{1}{z}$, it follows that $\dfrac{1}{z^2}f\left(\dfrac{1}{z}\right)=\dfrac{1}{z}$. Evidently, then,

$$\int_C f(z)\,dz=2\pi i \operatorname*{Res}_{z=0}\frac{1}{z^2}f\left(\frac{1}{z}\right)=2\pi i(1)=2\pi i.$$

4. Let C denote the circle $|z|=1$, taken counterclockwise.

(a) The Maclaurin series $e^z=\sum_{n=0}^{\infty}\dfrac{z^n}{n!}$ $(|z|<\infty)$ enables us to write

$$\int_C \exp\left(z+\frac{1}{z}\right)dz=\int_C e^z e^{1/z}\,dz=\int_C e^{1/z}\sum_{n=0}^{\infty}\frac{z^n}{n!}\,dz=\sum_{n=0}^{\infty}\frac{1}{n!}\int_C z^n\exp\left(\frac{1}{z}\right)dz.$$

(b) Referring to the Maclaurin series for e^z once again, let us write

$$z^n\exp\left(\frac{1}{z}\right)=z^n\sum_{k=0}^{\infty}\frac{1}{k!}\cdot\frac{1}{z^k}=\sum_{k=0}^{\infty}\frac{1}{k!}z^{n-k} \qquad (n=0,1,2,\ldots).$$

Now the $\dfrac{1}{z}$ in this series occurs when $n-k=-1$, or $k=n+1$. So, by the residue theorem,

$$\int_C z^n\exp\left(\frac{1}{z}\right)dz=2\pi i\frac{1}{(n+1)!} \qquad (n=0,1,2,\ldots).$$

The final result in part *(a)* thus reduces to

$$\int_C \exp\left(z+\frac{1}{z}\right)dz=2\pi i\sum_{n=0}^{\infty}\frac{1}{n!(n+1)!}.$$

5. We are given two polynomials

$$P(z) = a_0 + a_1 z + a_2 z^2 + \cdots + a_n z^n \qquad (a_n \neq 0)$$

and

$$Q(z) = b_0 + b_1 z + b_2 z^2 + \cdots + b_m z^m \qquad (b_m \neq 0),$$

where $m \geq n+2$.

It is straightforward to show that

$$\frac{1}{z^2} \cdot \frac{P(1/z)}{Q(1/z)} = \frac{a_0 z^{m-2} + a_1 z^{m-3} + a_2 z^{m-4} + \cdots + a_n z^{m-n-2}}{b_0 z^m + b_1 z^{m-1} + b_2 z^{m-2} + \cdots + b_m} \qquad (z \neq 0).$$

Observe that the numerator here is, in fact, a polynomial since $m-n-2 \geq 0$. Also, since $b_m \neq 0$, the quotient of these polynomials is represented by a series of the form $d_0 + d_1 z + d_2 z^2 + \cdots$. That is,

$$\frac{1}{z^2} \cdot \frac{P(1/z)}{Q(1/z)} = d_0 + d_1 z + d_2 z^2 + \cdots \qquad (0 < |z| < R_2);$$

and we see that $\frac{1}{z^2} \cdot \frac{P(1/z)}{Q(1/z)}$ has residue 0 $z = 0$.

Suppose now that all of the zeros of $Q(z)$ lie inside a simple closed contour C, and assume that C is positively oriented. Since $P(z)/Q(z)$ is analytic everywhere in the finite plane except at the zeros of $Q(z)$, it follows from the theorem in Sec. 64 and the residue just obtained that

$$\int_C \frac{P(z)}{Q(z)} dz = 2\pi i \operatorname*{Res}_{z=0} \left[\frac{1}{z^2} \cdot \frac{P(1/z)}{Q(1/z)} \right] = 2\pi i \cdot 0 = 0.$$

If C is negatively oriented, this result is still true since then

$$\int_C \frac{P(z)}{Q(z)} dz = -\int_{-C} \frac{P(z)}{Q(z)} dz = 0.$$

SECTION 65

1. *(a)* From the expansion

$$e^z = 1 + \frac{z}{1!} + \frac{z^2}{2!} + \frac{z^3}{3!} + \cdots \qquad (|z| < \infty),$$

we see that

$$z \exp\left(\frac{1}{z}\right) = z + 1 + \frac{1}{2!} \cdot \frac{1}{z} + \frac{1}{3!} \cdot \frac{1}{z^2} + \cdots \qquad (0 < |z| < \infty).$$

The principal part of $z\exp\left(\frac{1}{z}\right)$ at the isolated singular point $z=0$ is, then,

$$\frac{1}{2!}\cdot\frac{1}{z}+\frac{1}{3!}\cdot\frac{1}{z^2}+\cdots;$$

and $z=0$ is an essential singular point of that function.

(b) The isolated singular point of $\frac{z^2}{1+z}$ is at $z=-1$. Since the principal part at $z=-1$ involves powers of $z+1$, we begin by observing that

$$z^2=(z+1)^2-2z-1=(z+1)^2-2(z+1)+1.$$

This enables us to write

$$\frac{z^2}{1+z}=\frac{(z+1)^2-2(z+1)+1}{z+1}=(z+1)-2+\frac{1}{z+1}.$$

Since the principal part is $\frac{1}{z+1}$, the point $z=-1$ is a (simple) pole.

(c) The point $z=0$ is the isolated singular point of $\frac{\sin z}{z}$, and we can write

$$\frac{\sin z}{z}=\frac{1}{z}\left(z-\frac{z^3}{3!}+\frac{z^5}{5!}-\cdots\right)=1-\frac{z^2}{3!}+\frac{z^4}{5!}-\cdots \qquad (0<|z|<\infty).$$

The principal part here is evidently 0, and so $z=0$ is a removable singular point of the function $\frac{\sin z}{z}$.

(d) The isolated singular point of $\frac{\cos z}{z}$ is $z=0$. Since

$$\frac{\cos z}{z}=\frac{1}{z}\left(1-\frac{z^2}{2!}+\frac{z^4}{4!}-\cdots\right)=\frac{1}{z}-\frac{z}{2!}+\frac{z^3}{4!}-\cdots \qquad (0<|z|<\infty),$$

the principal part is $\frac{1}{z}$. This means that $z=0$ is a (simple) pole of $\frac{\cos z}{z}$.

(e) Upon writing $\frac{1}{(2-z)^3}=\frac{-1}{(z-2)^3}$, we find that the principal part of $\frac{1}{(2-z)^3}$ at its isolated singular point $z=2$ is simply the function itself. That point is evidently a pole (of order 3).

2. *(a)* The singular point is $z=0$. Since

$$\frac{1-\cosh z}{z^3}=\frac{1}{z^3}\left[1-\left(1+\frac{z^2}{2!}+\frac{z^4}{4!}+\frac{z^6}{6!}+\cdots\right)\right]=-\frac{1}{2!}\cdot\frac{1}{z}-\frac{z}{4!}-\frac{z^3}{6!}-\cdots$$

when $0<|z|<\infty$, we have $m=1$ and $B=-\frac{1}{2!}=-\frac{1}{2}$.

(b) Here the singular point is also $z=0$. Since

$$\frac{1-\exp(2z)}{z^4}=\frac{1}{z^4}\left[1-\left(1+\frac{2z}{1!}+\frac{2^2z^2}{2!}+\frac{2^3z^3}{3!}+\frac{2^4z^4}{4!}+\frac{2^5z^5}{5!}+\cdots\right)\right]$$

$$=-\frac{2}{1!}\cdot\frac{1}{z^3}-\frac{2^2}{2!}\cdot\frac{1}{z^2}-\frac{2^3}{3!}\cdot\frac{1}{z}-\frac{2^4}{4!}-\frac{2^5}{5!}z-\cdots$$

when $0<|z|<\infty$, we have $m=3$ and $B=-\frac{2^3}{3!}=-\frac{4}{3}$.

(c) The singular point of $\frac{\exp(2z)}{(z-1)^2}$ is $z=1$. The Taylor series

$$\exp(2z)=e^{2(z-1)}e^2=e^2\left[1+\frac{2(z-1)}{1!}+\frac{2^2(z-1)^2}{2!}+\frac{2^3(z-1)^3}{3!}+\cdots\right]\qquad(|z|<\infty)$$

enables us to write the Laurent series

$$\frac{\exp(2z)}{(z-1)^2}=e^2\left[\frac{1}{(z-1)^2}+\frac{2}{1!}\cdot\frac{1}{z-1}+\frac{2^2}{2!}+\frac{2^2}{3!}(z-1)+\cdots\right]\qquad(0<|z-1|<\infty).$$

Thus $m=2$ and $B=e^2\frac{2}{1!}=2e^2$.

3. Since f is analytic at z_0, it has a Taylor series representation

$$f(z)=f(z_0)+\frac{f'(z_0)}{1!}(z-z_0)+\frac{f''(z_0)}{2!}(z-z_0)^2+\cdots\qquad(|z-z_0|<R_0).$$

Let g be defined by means of the equation

$$g(z)=\frac{f(z)}{z-z_0}.$$

(a) Suppose that $f(z_0) \neq 0$. Then

$$g(z) = \frac{1}{z - z_0}\left[f(z_0) + \frac{f'(z_0)}{1!}(z - z_0) + \frac{f''(z_0)}{2!}(z - z_0)^2 + \cdots\right]$$

$$= \frac{f(z_0)}{z - z_0} + \frac{f'(z_0)}{1!} + \frac{f''(z_0)}{2!}(z - z_0) + \cdots \qquad (0 < |z - z_0| < R_0).$$

This shows that g has a simple pole at z_0, with residue $f(z_0)$.

(b) Suppose, on the other hand, that $f(z_0) = 0$. Then

$$g(z) = \frac{1}{z - z_0}\left[\frac{f'(z_0)}{1!}(z - z_0) + \frac{f''(z_0)}{2!}(z - z_0)^2 + \cdots\right]$$

$$= \frac{f'(z_0)}{1!} + \frac{f''(z_0)}{2!}(z - z_0) + \cdots \qquad (0 < |z - z_0| < R_0).$$

Since the principal part of g at z_0 is just 0, the point $z = 0$ is a removable singular point of g.

4. Write the function

$$f(z) = \frac{8a^3 z^2}{(z^2 + a^2)^3} \qquad (a > 0)$$

as

$$f(z) = \frac{\phi(z)}{(z - ai)^3} \quad \text{where} \quad \phi(z) = \frac{8a^3 z^2}{(z + ai)^3}.$$

Since the only singularity of $\phi(z)$ is at $z = -ai$, $\phi(z)$ has a Taylor series representation

$$\phi(z) = \phi(ai) + \frac{\phi'(ai)}{1!}(z - ai) + \frac{\phi''(ai)}{2!}(z - ai)^2 + \cdots \qquad (|z - ai| < 2a)$$

about $z = ai$. Thus

$$f(z) = \frac{1}{(z - ai)^3}\left[\phi(ai) + \frac{\phi'(ai)}{1!}(z - ai) + \frac{\phi''(ai)}{2!}(z - ai)^2 + \cdots\right] \qquad (0 < |z - ai| < 2a).$$

Now straightforward differentiation reveals that

$$\phi'(z) = \frac{16a^4 iz - 8a^3 z^2}{(z + ai)^4} \quad \text{and} \quad \phi''(z) = \frac{16a^3(z^2 - 4aiz - a^2)}{(z + ai)^5}.$$

Consequently,

$$\phi(ai) = -a^2 i, \quad \phi'(ai) = -\frac{a}{2}, \quad \text{and} \quad \phi''(ai) = -i.$$

This enables us to write

$$f(z) = \frac{1}{(z-ai)^3}\left[-a^2 i - \frac{a}{2}(z-ai) - \frac{i}{2}(z-ai)^2 + \cdots\right] \qquad (0 < |z-ai| < 2a).$$

The principal part of f at the point $z = ai$ is, then,

$$-\frac{i/2}{z-ai} - \frac{a/2}{(z-ai)^2} - \frac{a^2 i}{(z-ai)^3}.$$

SECTION 67

1. *(a)* The function $f(z) = \dfrac{z^2+2}{z-1}$ has an isolated singular point at $z = 1$. Writing $f(z) = \dfrac{\phi(z)}{z-1}$, where $\phi(z) = z^2 + 2$, and observing that $\phi(z)$ is analytic and nonzero at $z = 1$, we see that $z = 1$ is a pole of order $m = 1$ and that the residue there is $B = \phi(1) = 3$.

(b) If we write

$$f(z) = \left(\frac{z}{2z+1}\right)^3 = \frac{\phi(z)}{\left[z - \left(-\frac{1}{2}\right)\right]^3}, \quad \text{where} \quad \phi(z) = \frac{z^3}{8},$$

we see that $z = -\dfrac{1}{2}$ is a singular point of f. Since $\phi(z)$ is analytic and nonzero at that point, f has a pole of order $m = 3$ there. The residue is

$$B = \frac{\phi''(-1/2)}{2!} = -\frac{3}{16}.$$

(c) The function

$$\frac{\exp z}{z^2 + \pi^2} = \frac{\exp z}{(z - \pi i)(z + \pi i)}$$

has poles of order $m = 1$ at the two points $z = \pm \pi i$. The residue at $z = \pi i$ is

$$B_1 = \frac{\exp \pi i}{2\pi i} = \frac{-1}{2\pi i} = \frac{i}{2\pi},$$

and the one at $z = -\pi i$ is

$$B_2 = \frac{\exp(-\pi i)}{-2\pi i} = \frac{-1}{-2\pi i} = -\frac{i}{2\pi}.$$

2. *(a)* Write the function $f(z)=\dfrac{z^{1/4}}{z+1}$ $(|z|>0, 0<\arg z<2\pi)$ as

$$f(z)=\frac{\phi(z)}{z+1}, \quad \text{where} \quad \phi(z)=z^{1/4}=e^{\frac{1}{4}\log z} \quad (|z|>0, 0<\arg z<2\pi).$$

The function $\phi(z)$ is analytic throughout its domain of definition, indicated in the figure below.

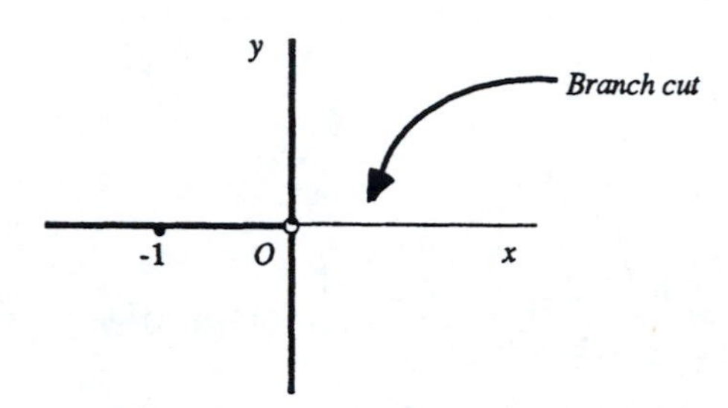

Also,

$$\phi(-1)=(-1)^{1/4}=e^{\frac{1}{4}\log(-1)}=e^{\frac{1}{4}(\ln 1+i\pi)}=e^{i\pi/4}=\cos\frac{\pi}{4}+i\sin\frac{\pi}{4}=\frac{1+i}{\sqrt{2}}\neq 0.$$

This shows that the function *f* has a pole of order $m=1$ at $z=-1$, the residue there being

$$B=\phi(-1)=\frac{1+i}{\sqrt{2}}.$$

(b) Write the function $f(z)=\dfrac{\operatorname{Log} z}{(z^2+1)^2}$ as

$$f(z)=\frac{\phi(z)}{(z-i)^2} \quad \text{where} \quad \phi(z)=\frac{\operatorname{Log} z}{(z+i)^2}.$$

From this, it is clear that $f(z)$ has a pole of order $m=2$ at $z=i$. Straightforward differentiation then reveals that

$$\operatorname*{Res}_{z=i}\frac{\operatorname{Log} z}{(z^2+1)^2}=\phi'(i)=\frac{\pi+2i}{8}.$$

(*c*) Write the function

$$f(z) = \frac{z^{1/2}}{(z^2+1)^2} \qquad (|z| > 0, 0 < \arg z < 2\pi)$$

as

$$f(z) = \frac{\phi(z)}{(z-i)^2} \quad \text{where} \quad \phi(z) = \frac{z^{1/2}}{(z+i)^2}.$$

Since

$$\phi'(z) = \frac{(z+i)z^{-1/2} - 4z^{1/2}}{2(z+i)^3}$$

and

$$i^{-1/2} = e^{-i\pi/4} = \frac{1}{\sqrt{2}} - \frac{i}{\sqrt{2}}, \qquad i^{1/2} = e^{i\pi/4} = \frac{1}{\sqrt{2}} + \frac{i}{\sqrt{2}},$$

$$\operatorname*{Res}_{z=i} \frac{z^{1/2}}{(z^2+1)^2} = \phi'(i) = \frac{1-i}{8\sqrt{2}}.$$

3. (*a*) We wish to evaluate the integral

$$\int_C \frac{3z^3+2}{(z-1)(z^2+9)} dz,$$

where C is the circle $|z-2| = 2$, taken in the counterclockwise direction. That circle and the singularities $z = 1, \pm 3i$ of the integrand are shown in the figure just below.

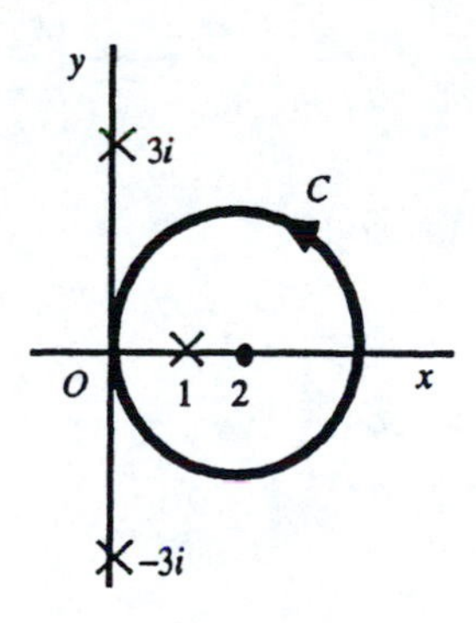

Observe that the point $z = 1$, which is the only singularity inside C, is a simple pole of the integrand and that

$$\operatorname*{Res}_{z=1} \frac{3z^3+2}{(z-1)(z^2+9)} = \left.\frac{3z^3+2}{z^2+9}\right]_{z=1} = \frac{1}{2}.$$

According to the residue theorem, then,

$$\int_C \frac{3z^3+2}{(z-1)(z^2+9)} dz = 2\pi i\left(\frac{1}{2}\right) = \pi i.$$

(b) Let us redo part *(a)* when C is changed to be the positively oriented circle $|z|=4$, shown in the figure below.

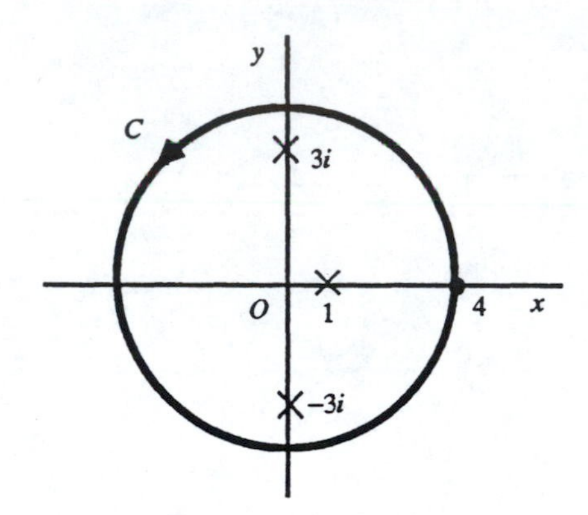

In this case, all three singularities $z=1, \pm 3i$ of the integrand are interior to C. We already know from part *(a)* that

$$\operatorname*{Res}_{z=1} \frac{3z^3+2}{(z-1)(z^2+9)} = \frac{1}{2}.$$

It is, moreover, straightforward to show that

$$\operatorname*{Res}_{z=3i} \frac{3z^3+2}{(z-1)(z^2+9)} = \left.\frac{3z^3+2}{(z-1)(z+3i)}\right]_{z=3i} = \frac{15+49i}{12}$$

and

$$\operatorname*{Res}_{z=-3i} \frac{3z^3+2}{(z-1)(z^2+9)} = \left.\frac{3z^3+2}{(z-1)(z-3i)}\right]_{z=-3i} = \frac{15-49i}{12}.$$

The residue theorem now tells us that

$$\int_C \frac{3z^3+2}{(z-1)(z^2+9)}\,dz = 2\pi i\left(\frac{1}{2} + \frac{15+49i}{12} + \frac{15-49i}{12}\right) = 6\pi i.$$

4. *(a)* Let C denote the positively oriented circle $|z|=2$, and note that the integrand of the integral $\int_C \frac{dz}{z^3(z+4)}$ has singularities at $z=0$ and $z=-4$. (See the figure below.)

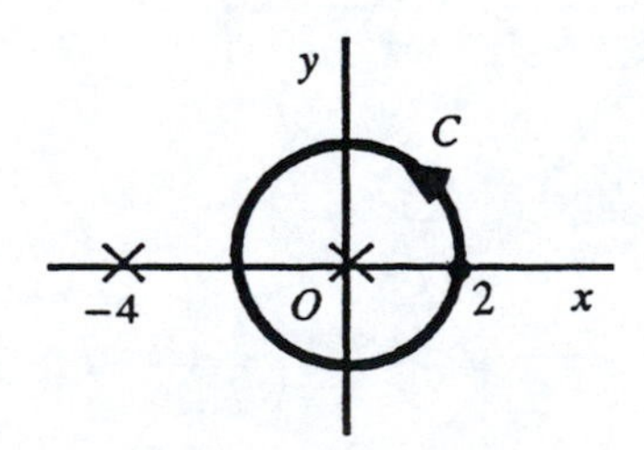

To find the residue of the integrand at $z=0$, we recall the expansion

$$\frac{1}{1-z}=\sum_{n=0}^{\infty} z^n \qquad (|z|<1)$$

and write

$$\frac{1}{z^3(z+4)}=\frac{1}{4z^3}\left[\frac{1}{1+(z/4)}\right]=\frac{1}{4z^3}\sum_{n=0}^{\infty}\left(-\frac{z}{4}\right)^n=\sum_{n=0}^{\infty}\frac{(-1)^n}{4^{n+1}}z^{n-3} \qquad (0<|z|<4).$$

Now the coefficient of $\frac{1}{z}$ here occurs when $n=2$, and we see that

$$\operatorname*{Res}_{z=0}\frac{1}{z^3(z+4)}=\frac{1}{64}.$$

Consequently,

$$\int_C \frac{dz}{z^3(z+4)}=2\pi i\left(\frac{1}{64}\right)=\frac{\pi i}{32}.$$

(b) Let us replace the path C in part *(a)* by the positively oriented circle $|z+2|=3$, centered at -2 and with radius 3. It is shown below.

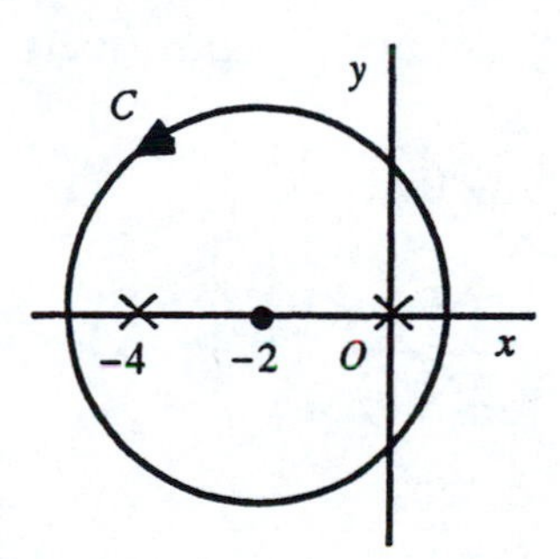

We already know from part *(a)* that

$$\operatorname*{Res}_{z=0}\frac{1}{z^3(z+4)}=\frac{1}{64}.$$

To find the residue at -4, we write

$$\frac{1}{z^3(z+4)}=\frac{\phi(z)}{z-(-4)}, \quad \text{where} \quad \phi(z)=\frac{1}{z^3}.$$

This tells us that $z=-4$ is a simple pole of the integrand and that the residue there is $\phi(-4)=-1/64$. Consequently,

$$\int_C \frac{dz}{z^3(z+4)}=2\pi i\left(\frac{1}{64}-\frac{1}{64}\right)=0.$$

5. Let us evaluate the integral $\int_C \frac{\cosh \pi z\, dz}{z(z^2+1)}$, where C is the positively oriented circle $|z|=2$. All three isolated singularities $z=0,\pm i$ of the integrand are interior to C. The desired residues are

$$\operatorname*{Res}_{z=0} \frac{\cosh \pi z}{z(z^2+1)} = \left.\frac{\cosh \pi z}{z^2+1}\right]_{z=0} = 1,$$

$$\operatorname*{Res}_{z=i} \frac{\cosh \pi z}{z(z^2+1)} = \left.\frac{\cosh \pi z}{z(z+i)}\right]_{z=i} = \frac{1}{2},$$

and

$$\operatorname*{Res}_{z=-i} \frac{\cosh \pi z}{z(z^2+1)} = \left.\frac{\cosh \pi z}{z(z-i)}\right]_{z=-i} = \frac{1}{2}.$$

Consequently,

$$\int_C \frac{\cosh \pi z\, dz}{z(z^2+1)} = 2\pi i\left(1+\frac{1}{2}+\frac{1}{2}\right) = 4\pi i.$$

6. In each part of this problem, C denotes the positively oriented circle $|z|=3$.

(a) It is straightforward to show that

$$\text{if } f(z) = \frac{(3z+2)^2}{z(z-1)(2z+5)}, \quad \text{then} \quad \frac{1}{z^2} f\left(\frac{1}{z}\right) = \frac{(3+2z)^2}{z(1-z)(2+5z)}.$$

This function $\frac{1}{z^2} f\left(\frac{1}{z}\right)$ has a simple pole at $z=0$, and

$$\int_C \frac{(3z+2)^2}{z(z-1)(2z+5)}\, dz = 2\pi i \operatorname*{Res}_{z=0}\left[\frac{1}{z^2} f\left(\frac{1}{z}\right)\right] = 2\pi i\left(\frac{9}{2}\right) = 9\pi i.$$

(b) Likewise,

$$\text{if } f(z) = \frac{z^3(1-3z)}{(1+z)(1+2z^4)}, \quad \text{then} \quad \frac{1}{z^2} f\left(\frac{1}{z}\right) = \frac{z-3}{z(z+1)(z^4+2)}.$$

The function $\frac{1}{z^2} f\left(\frac{1}{z}\right)$ has a simple pole at $z=0$, and we find here that

$$\int_C \frac{z^3(1-3z)}{(1+z)(1+2z^4)}\, dz = 2\pi i \operatorname*{Res}_{z=0}\left[\frac{1}{z^2} f\left(\frac{1}{z}\right)\right] = 2\pi i\left(-\frac{3}{2}\right) = -3\pi i.$$

(c) Finally,

$$\text{if } f(z)=\frac{z^3e^{1/z}}{1+z^3}, \quad \text{then} \quad \frac{1}{z^2}f\left(\frac{1}{z}\right)=\frac{e^z}{z^2(1+z^3)}.$$

The point $z=0$ is a pole of order 2 of $\frac{1}{z^2}f\left(\frac{1}{z}\right)$. The residue is $\phi'(0)$, where

$$\varphi(z)=\frac{e^z}{1+z^3}.$$

Since

$$\phi'(z)=\frac{(1+z^3)e^z-e^z3z^2}{(1+z^3)^2},$$

the value of $\phi'(0)$ is 1. So

$$\int_C \frac{z^3e^{1/z}}{1+z^3}\,dz=2\pi i \operatorname*{Res}_{z=0}\left[\frac{1}{z^2}f\left(\frac{1}{z}\right)\right]=2\pi i(1)=2\pi i.$$

SECTION 69

1. *(a)* Write

$$\csc z=\frac{1}{\sin z}=\frac{p(z)}{q(z)}, \quad \text{where} \quad p(z)=1 \text{ and } q(z)=\sin z.$$

Since

$$p(0)=1\neq 0, \quad q(0)=\sin 0=0, \quad \text{and} \quad q'(0)=\cos 0=1\neq 0,$$

$z=0$ must be a simple pole of $\csc z$, with residue

$$\frac{p(0)}{q'(0)}=\frac{1}{1}=1.$$

(b) From Exercise 2, Sec. 61, we know that

$$\csc z=\frac{1}{z}+\frac{1}{3!}z+\left[\frac{1}{(3!)^2}-\frac{1}{5!}\right]z^3+\cdots \qquad (0<|z|<\pi).$$

Since the coefficient of $\frac{1}{z}$ here is 1, it follows that $z=0$ is a simple pole of $\csc z$, the residue being 1.

2. *(a)* Write

$$\frac{z-\sinh z}{z^2\sinh z}=\frac{p(z)}{q(z)}, \quad \text{where} \quad p(z)=z-\sinh z \text{ and } q(z)=z^2\sinh z.$$

Sincc

$$p(\pi i)=\pi i\neq 0, \quad q(\pi i)=0, \quad \text{and} \quad q'(\pi i)=\pi^2\neq 0,$$

it follows that

$$\operatorname*{Res}_{z=\pi i}\frac{z-\sinh z}{z^2\sinh z}=\frac{p(\pi i)}{q'(\pi i)}=\frac{\pi i}{\pi^2}=\frac{i}{\pi}.$$

(b) Write

$$\frac{\exp(zt)}{\sinh z}=\frac{p(z)}{q(z)}, \quad \text{where} \quad p(z)=\exp(zt) \text{ and } q(z)=\sinh z.$$

It is easy to see that

$$\operatorname*{Res}_{z=\pi i}\frac{\exp(zt)}{\sinh z}=\frac{p(\pi i)}{q'(\pi i)}=-\exp(i\pi t) \quad \text{and} \quad \operatorname*{Res}_{z=-\pi i}\frac{\exp(zt)}{\sinh z}=\frac{p(-\pi i)}{q'(-\pi i)}=-\exp(-i\pi t).$$

Evidently, then,

$$\operatorname*{Res}_{z=\pi i}\frac{\exp(zt)}{\sinh z}+\operatorname*{Res}_{z=-\pi i}\frac{\exp(zt)}{\sinh z}=-2\,\frac{\exp(i\pi t)+\exp(-i\pi t)}{2}=-2\cos\pi t.$$

3. *(a)* Write

$$f(z)=\frac{p(z)}{q(z)}, \quad \text{where} \quad p(z)=z \text{ and } q(z)=\cos z.$$

Observe that

$$q\left(\frac{\pi}{2}+n\pi\right)=0 \qquad (n=0,\pm1,\pm2,\ldots).$$

Also, for the stated values of n,

$$p\left(\frac{\pi}{2}+n\pi\right)=\frac{\pi}{2}+n\pi\neq 0 \quad \text{and} \quad q'\left(\frac{\pi}{2}+n\pi\right)=-\sin\left(\frac{\pi}{2}+n\pi\right)=(-1)^{n+1}\neq 0.$$

So the function $f(z)=\dfrac{z}{\cos z}$ has poles of order $m=1$ at each of the points

$$z_n=\frac{\pi}{2}+n\pi \qquad (n=0,\pm1,\pm2,\ldots).$$

The corresponding residues are

$$B=\frac{p(z_n)}{q'(z_n)}=(-1)^{n+1}z_n.$$

(b) Write

$$\tanh z=\frac{p(z)}{q(z)}, \quad \text{where} \quad p(z)=\sinh z \text{ and } q(z)=\cosh z.$$

Both p and q are entire, and the zeros of q are (Sec. 34)

$$z=\left(\frac{\pi}{2}+n\pi\right)i \qquad (n=0,\pm1,\pm2,\ldots)$$

In addition to the fact that $q\left(\left(\frac{\pi}{2}+n\pi\right)i\right)=0$, we see that

$$p\left(\left(\frac{\pi}{2}+n\pi\right)i\right)=\sinh\left(\frac{\pi}{2}i+n\pi i\right)=i\cos n\pi=i(-1)^n\neq 0$$

and

$$q'\left(\left(\frac{\pi}{2}+n\pi\right)i\right)=\sinh\left(\frac{\pi}{2}i+n\pi i\right)=i(-1)^n\neq 0.$$

So the points $z=\left(\frac{\pi}{2}+n\pi\right)i$ $(n=0,\pm1,\pm2,\ldots)$ are poles of order $m=1$ of $\tanh z$, the residue in each case being

$$B=\frac{p\left(\left(\frac{\pi}{2}+n\pi\right)i\right)}{q'\left(\left(\frac{\pi}{2}+n\pi\right)i\right)}=\frac{i(-1)^n}{i(-1)^n}=1.$$

4. Let C be the positively oriented circle $|z| = 2$, shown just below.

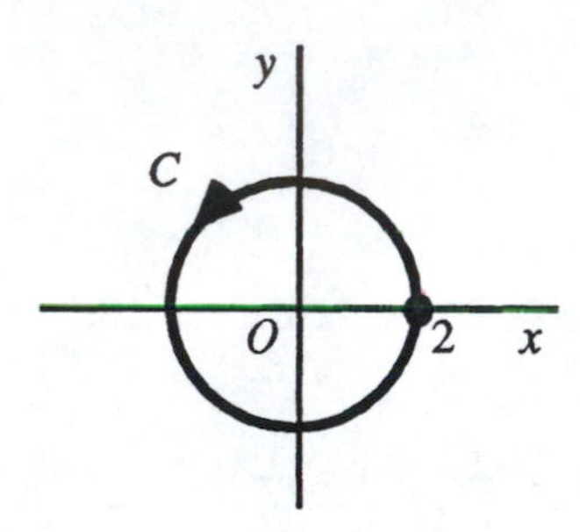

(a) To evaluate the integral $\int_C \tan z\,dz$, we write the integrand as

$$\tan z = \frac{p(z)}{q(z)}, \quad \text{where} \quad p(z) = \sin z \ \text{ and } \ q(z) = \cos z,$$

and recall that the zeros of $\cos z$ are $z = \frac{\pi}{2} + n\pi \ (n = 0, \pm 1, \pm 2, \ldots)$. Only two of those zeros, namely $z = \pm \pi/2$, are interior to C, and they are the isolated singularities of $\tan z$ interior to C. Observe that

$$\operatorname*{Res}_{z=\pi/2} \tan z = \frac{p(\pi/2)}{q'(\pi/2)} = -1 \quad \text{and} \quad \operatorname*{Res}_{z=-\pi/2} \tan z = \frac{p(-\pi/2)}{q'(-\pi/2)} = -1.$$

Hence

$$\int_C \tan z\,dz = 2\pi i(-1-1) = -4\pi i.$$

(b) The problem here is to evaluate the integral $\int_C \frac{dz}{\sinh 2z}$. To do this, we write the integrand as

$$\frac{1}{\sinh 2z} = \frac{p(z)}{q(z)}, \quad \text{where} \quad p(z) = 1 \ \text{ and } \ q(z) = \sinh 2z.$$

Now $\sinh 2z = 0$ when $2z = n\pi i \ (n = 0, \pm 1, \pm 2, \ldots)$, or when

$$z = \frac{n\pi i}{2} \qquad (n = 0, \pm 1, \pm 2, \ldots).$$

Three of these zeros of $\sinh 2z$, namely 0 and $\pm\frac{\pi i}{2}$, are inside C and are the isolated singularities of the integrand that need to be considered here. It is straightforward to show that

$$\operatorname*{Res}_{z=0} \frac{1}{\sinh 2z} = \frac{p(0)}{q'(0)} = \frac{1}{2\cosh 0} = \frac{1}{2},$$

$$\operatorname*{Res}_{z=\pi i/2} \frac{1}{\sinh 2z} = \frac{p(\pi i/2)}{q'(\pi i/2)} = \frac{1}{2\cosh(\pi i)} = \frac{1}{2\cos\pi} = -\frac{1}{2},$$

and

$$\operatorname*{Res}_{z=-\pi i/2} \frac{1}{\sinh 2z} = \frac{p(-\pi i/2)}{q'(-\pi i/2)} = \frac{1}{2\cosh(-\pi i)} = \frac{1}{2\cos(-\pi)} = -\frac{1}{2}.$$

Thus

$$\int_C \frac{dz}{\sinh 2z} = 2\pi i\left(\frac{1}{2} - \frac{1}{2} - \frac{1}{2}\right) = -\pi i.$$

5. The simple closed contour C_N is as shown in the figure below.

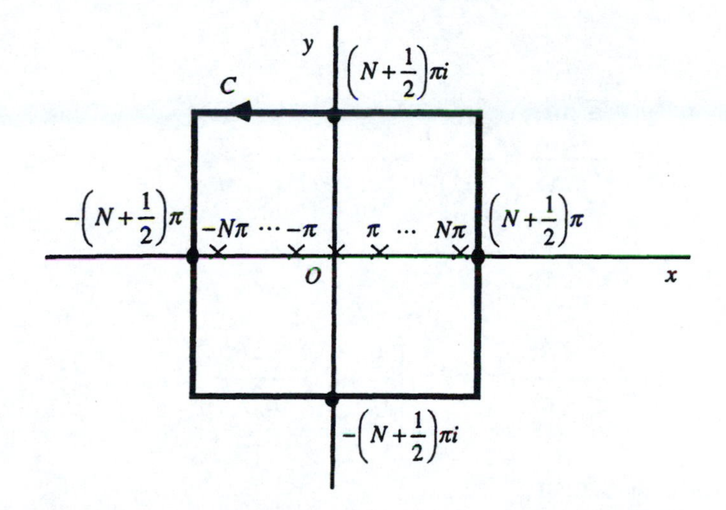

Within C_N, the function $\dfrac{1}{z^2 \sin z}$ has isolated singularities at

$$z = 0 \quad \text{and} \quad z = \pm n\pi \ (n = 1, 2, \ldots, N).$$

To find the residue at $z = 0$, we recall the Laurent series for $\csc z$ that was found in Exercise 2, Sec. 61, and write

$$\frac{1}{z^2 \sin z} = \frac{1}{z^2}\csc z = \frac{1}{z^2}\left\{\frac{1}{z} + \frac{1}{3!}z + \left[\frac{1}{(3!)^2} - \frac{1}{5!}\right]z^3 + \cdots\right\}$$

$$= \frac{1}{z^3} + \frac{1}{6}\cdot\frac{1}{z} + \left[\frac{1}{(3!)^2} - \frac{1}{5!}\right]z + \cdots \qquad (0 < |z| < \pi).$$

This tells us that $\dfrac{1}{z^2 \sin z}$ has a pole of order 3 at $z = 0$ and that

$$\operatorname*{Res}_{z=0} \frac{1}{z^2 \sin z} = \frac{1}{6}.$$

As for the points $z = \pm n\pi \ (n = 1, 2, \ldots, N)$, write

$$\frac{1}{z^2 \sin z} = \frac{p(z)}{q(z)}, \quad \text{where} \quad p(z) = 1 \text{ and } q(z) = z^2 \sin z.$$

Since

$$p(\pm n\pi) = 1 \neq 0, \quad q(\pm n\pi) = 0, \quad \text{and} \quad q'(\pm n\pi) = n^2\pi^2 \cos n\pi = (-1)^n n^2 \pi^2 \neq 0,$$

it follows that

$$\operatorname*{Res}_{z=\pm n\pi} \frac{1}{z^2 \sin z} = \frac{1}{(-1)^n n^2 \pi^2} \cdot \frac{(-1)^n}{(-1)^n} = \frac{(-1)^n}{n^2 \pi^2}.$$

So, by the residue theorem,

$$\int_{C_N} \frac{dz}{z^2 \sin z} dz = 2\pi i \left[\frac{1}{6} + 2 \sum_{n=1}^{N} \frac{(-1)^n}{n^2 \pi^2} \right].$$

Rewriting this equation in the form

$$\sum_{n=1}^{N} \frac{(-1)^{n+1}}{n^2} = \frac{\pi^2}{12} - \frac{\pi}{4i} \int_{C_N} \frac{dz}{z^2 \sin z}$$

and recalling from Exercise 7, Sec. 41, that the value of the integral here tends to zero as N tends to infinity, we arrive at the desired summation formula:

$$\sum_{n=1}^{\infty} \frac{(-1)^{n+1}}{n^2} = \frac{\pi^2}{12}.$$

6. The path C here is the positively oriented boundary of the rectangle with vertices at the points ± 2 and $\pm 2 + i$. The problem is to evaluate the integral

$$\int_C \frac{dz}{(z^2 - 1)^2 + 3}.$$

The isolated singularities of the integrand are the zeros of the polynomial

$$q(z) = (z^2 - 1)^2 + 3.$$

Setting this polynomial equal to zero and solving for z^2, we find that any zero z of $q(z)$ has the property $z^2 = 1 \pm \sqrt{3}i$. It is straightforward to find the two square roots of $1+\sqrt{3}i$ and also the two square roots of $1-\sqrt{3}i$. These are the four zeros of $q(z)$. Only two of those zeros,

$$z_0 = \sqrt{2}e^{i\pi/6} = \frac{\sqrt{3}+i}{\sqrt{2}} \quad \text{and} \quad -\overline{z}_0 = -\sqrt{2}e^{-i\pi/6} = \frac{-\sqrt{3}+i}{\sqrt{2}},$$

lie inside C. They are shown in the figure below.

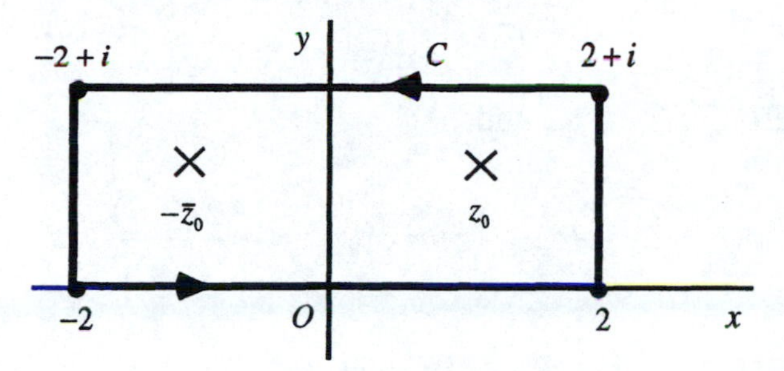

To find the residues at z_0 and $-\overline{z}_0$, we write the integrand of the integral to be evaluated as

$$\frac{1}{(z^2-1)^2+3} = \frac{p(z)}{q(z)}, \quad \text{where} \quad p(z) = 1 \text{ and } q(z) = (z^2-1)^2 + 3.$$

This polynomial $q(z)$ is, of course, the same $q(z)$ as above; hence $q(z_0) = 0$. Note, too, that p and q are analytic at z_0 and that $p(z_0) \neq 0$. Finally, it is straightforward to show that $q'(z) = 4z(z^2 - 1)$ and hence that

$$q'(z_0) = 4z_0(z_0^2 - 1) = -2\sqrt{6} + 6\sqrt{2}i \neq 0.$$

We may conclude, then, that z_0 is a simple pole of the integrand, with residue

$$\frac{p(z_0)}{q'(z_0)} = \frac{1}{-2\sqrt{6}+6\sqrt{2}i}.$$

Similar results are to be found at the singular point $-\overline{z}_0$. To be specific, it is easy to see that

$$q'(-\overline{z}_0) = -q'(\overline{z}_0) = -\overline{q'(z_0)} = 2\sqrt{6} + 6\sqrt{2}i \neq 0,$$

the residue of the integrand at $-\overline{z}_0$ being

$$\frac{p(-\overline{z}_0)}{q'(-\overline{z}_0)} = \frac{1}{2\sqrt{6}+6\sqrt{2}i}.$$

Finally, by the residue theorem,

$$\int_C \frac{dz}{(z^2-1)^2+3} = 2\pi i\left(\frac{1}{-2\sqrt{6}+6\sqrt{2}i} + \frac{1}{2\sqrt{6}+6\sqrt{2}i}\right) = \frac{\pi}{2\sqrt{2}}.$$

7. We are given that $f(z) = 1/[q(z)]^2$, where q is analytic at z_0, $q(z_0)=0$, and $q'(z_0) \neq 0$. These conditions on q tell us that q has a zero of order $m=1$ at z_0. Hence $q(z)=(z-z_0)g(z)$, where g is a function that is analytic and nonzero at z_0; and this enables us to write

$$f(z) = \frac{\phi(z)}{(z-z_0)^2}, \quad \text{where} \quad \phi(z) = \frac{1}{[g(z)]^2}.$$

So f has a pole of order 2 at z_0, and

$$\operatorname*{Res}_{z=z_0} f(z) = \phi'(z_0) = -\frac{2g'(z_0)}{[g(z_0)]^3}.$$

But, since $q(z)=(z-z_0)g(z)$, we know that

$$q'(z) = (z-z_0)g'(z) + g(z) \quad \text{and} \quad q''(z) = (z-z_0)g''(z) + 2g'(z).$$

Then, by setting $z = z_0$ in these last two equations, we find that

$$q'(z_0) = g(z_0) \quad \text{and} \quad q''(z_0) = 2g'(z_0).$$

Consequently, our expression for the residue of f at z_0 can be put in the desired form:

$$\operatorname*{Res}_{z=0} f(z) = -\frac{q''(z_0)}{[q'(z_0)]^3}.$$

8. *(a)* To find the residue of the function $\csc^2 z$ at $z=0$, we write

$$\csc^2 z = \frac{1}{[q(z)]^2}, \quad \text{where} \quad q(z) = \sin z.$$

Since q is entire, $q(0)=0$, and $q'(0)=1\neq 0$, the result in Exercise 7 tells us that

$$\operatorname*{Res}_{z=0} \csc^2 z = -\frac{q''(0)}{[q'(0)]^3} = 0.$$

(b) The residue of the function $\dfrac{1}{(z+z^2)^2}$ at $z=0$ can be obtained by writing

$$\frac{1}{(z+z^2)^2}=\frac{1}{[q(z)]^2}, \quad \text{where} \quad q(z)=z+z^2.$$

Inasmuch as q is entire, $q(0)=0$, and $q'(0)=1\neq 0$, we know from Exercise 7 that

$$\operatorname*{Res}_{z=0}\frac{1}{(z+z^2)^2}=-\frac{q''(0)}{[q'(0)]^3}=-2.$$

Chapter 7

SECTION 72

1. To evaluate the integral $\int_0^\infty \frac{dx}{x^2+1}$, we integrate the function $f(z)=\frac{1}{z^2+1}$ around the simple closed contour shown below, where $R>1$.

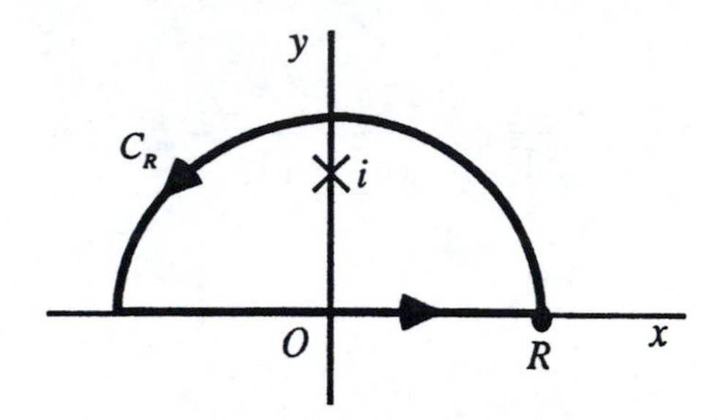

We see that

$$\int_{-R}^{R} \frac{dx}{x^2+1} + \int_{C_R} \frac{dz}{z^2+1} = 2\pi i B,$$

where

$$B = \operatorname*{Res}_{z=i} \frac{1}{z^2+1} = \operatorname*{Res}_{z=i} \frac{1}{(z-i)(z+i)} = \frac{1}{z+i}\bigg]_{z=i} = \frac{1}{2i}.$$

Thus

$$\int_{-R}^{R} \frac{dx}{x^2+1} = \pi - \int_{C_R} \frac{dz}{z^2+1}.$$

Now if z is a point on C_R,

$$|z^2+1| \geq ||z|^2-1| = R^2-1;$$

and so

$$\left|\int_{C_R} \frac{dz}{z^2+1}\right| \leq \frac{\pi R}{R^2-1} = \frac{\dfrac{\pi}{R}}{1-\dfrac{1}{R^2}} \to 0 \quad \text{as} \quad R\to\infty.$$

Finally, then

$$\int_{-\infty}^{\infty} \frac{dx}{x^2+1} = \pi, \quad \text{or} \quad \int_0^\infty \frac{dx}{x^2+1} = \frac{\pi}{2}.$$

2. The integral $\int_0^\infty \frac{dx}{(x^2+1)^2}$ can be evaluated using the function $f(z)=\frac{1}{(z^2+1)^2}$ and the same simple closed contour as in Exercise 1. Here

$$\int_{-R}^{R} \frac{dx}{(x^2+1)^2} + \int_{C_R} \frac{dz}{(z^2+1)^2} = 2\pi i B,$$

where $B = \operatorname*{Res}_{z=i} \frac{1}{(z^2+1)^2}$. Since

$$\frac{1}{(z^2+1)^2} = \frac{\phi(z)}{(z-i)^2}, \quad \text{where} \quad \phi(z) = \frac{1}{(z+i)^2},$$

we readily find that $B = \phi'(i) = \frac{1}{4i}$, and so

$$\int_{-R}^{R} \frac{dx}{(x^2+1)^2} = \frac{\pi}{2} - \int_{C_R} \frac{dz}{(z^2+1)^2}.$$

If z is a point on C_R, we know from Exercise 1 that

$$|z^2+1| \geq R^2 - 1;$$

thus

$$\left| \int_{C_R} \frac{dz}{(z^2+1)^2} \right| \leq \frac{\pi R}{(R^2-1)^2} = \frac{\dfrac{\pi}{R^3}}{\left(1-\dfrac{1}{R^2}\right)^2} \to 0 \quad \text{as} \quad R \to \infty.$$

The desired result is, then,

$$\int_{-\infty}^{\infty} \frac{dx}{(x^2+1)^2} = \frac{\pi}{2}, \quad \text{or} \quad \int_0^\infty \frac{dx}{(x^2+1)^2} = \frac{\pi}{4}.$$

3. We begin the evaluation of $\int_0^\infty \frac{dx}{x^4+1}$ by finding the zeros of the polynomial z^4+1, which are the fourth roots of -1, and noting that two of them are below the real axis. In fact, if we consider the simple closed contour shown below, where $R>1$, that contour encloses only the two roots

$$z_1 = e^{i\pi/4} = \frac{1}{\sqrt{2}} + \frac{i}{\sqrt{2}}$$

and

$$z_2 = e^{i3\pi/4} = e^{i\pi/4}e^{i\pi/2} = \left(\frac{1}{\sqrt{2}} + \frac{i}{\sqrt{2}}\right)i = -\frac{1}{\sqrt{2}} + \frac{i}{\sqrt{2}}.$$

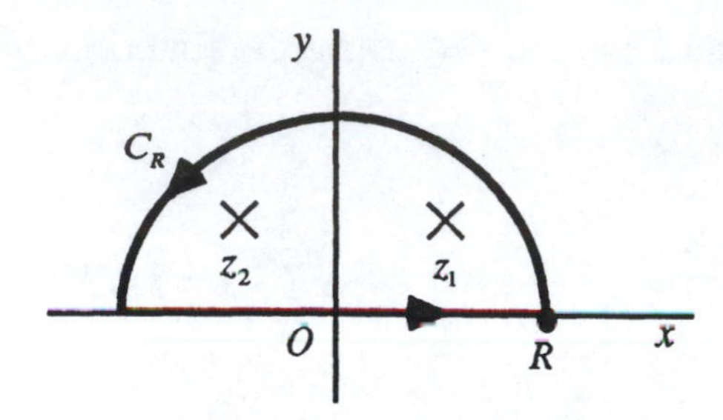

Now

$$\int_{-R}^{R} \frac{dx}{x^4+1} + \int_{C_R} \frac{dz}{z^4+1} = 2\pi i(B_1 + B_2),$$

where

$$B_1 = \operatorname*{Res}_{z=z_1} \frac{1}{z^4+1} \quad \text{and} \quad B_2 = \operatorname*{Res}_{z=z_2} \frac{1}{z^4+1}.$$

The method of Theorem 2 in Sec. 69 tells us that z_1 and z_2 are simple poles of $\dfrac{1}{z^4+1}$ and that

$$B_1 = \frac{1}{4z_1^3} \cdot \frac{z_1}{z_1} = -\frac{z_1}{4} \quad \text{and} \quad B_2 = \frac{1}{4z_2^3} \cdot \frac{z_2}{z_2} = -\frac{z_2}{4},$$

since $z_1^4 = -1$ and $z_2^4 = -1$. Furthermore,

$$B_1 + B_2 = -\frac{1}{4}(z_1 + z_2) = -\frac{1}{4}\left[\left(\frac{1}{\sqrt{2}} + \frac{i}{\sqrt{2}}\right) + \left(-\frac{1}{\sqrt{2}} + \frac{i}{\sqrt{2}}\right)\right] = -\frac{i}{2\sqrt{2}}.$$

Hence

$$\int_{-R}^{R} \frac{dx}{x^4+1} = \frac{\pi}{\sqrt{2}} - \int_{C_R} \frac{dz}{z^4+1}.$$

Since

$$\left|\int_{C_R} \frac{dz}{z^4+1}\right| \le \frac{\pi R}{R^4 - 1} \to 0 \text{ as } R \to \infty,$$

we have

$$\int_{-\infty}^{\infty} \frac{dx}{x^4+1} = \frac{\pi}{\sqrt{2}}, \quad \text{or} \quad \int_{0}^{\infty} \frac{dx}{x^4+1} = \frac{\pi}{2\sqrt{2}}.$$

4. We wish to evaluate the integral $\int_0^\infty \frac{x^2\,dx}{(x^2+1)(x^2+4)}$. We use the simple closed contour shown below, where $R > 2$.

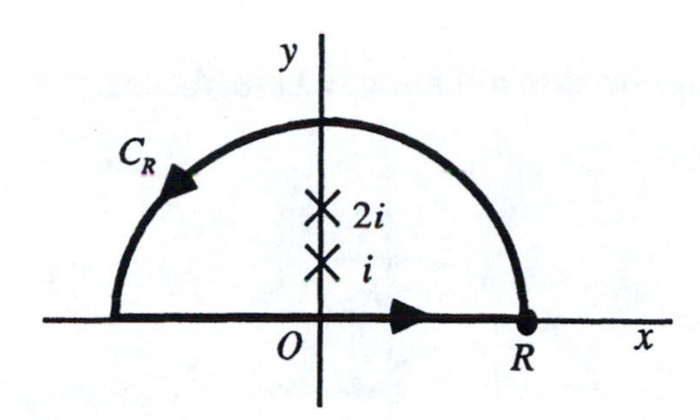

We must find the residues of the function $f(z) = \frac{z^2}{(z^2+1)(z^2+4)}$ at its simple poles $z = i$ and $z = 2i$. They are

$$B_1 = \operatorname*{Res}_{z=i} f(z) = \left.\frac{z^2}{(z+i)(z^2+4)}\right]_{z=i} = -\frac{1}{6i}$$

and

$$B_2 = \operatorname*{Res}_{z=2i} f(z) = \left.\frac{z^2}{(z^2+1)(z+2i)}\right]_{z=2i} = \frac{1}{3i}.$$

Thus

$$\int_{-R}^{R} \frac{x^2\,dx}{(x^2+1)(x^2+4)} + \int_{C_R} \frac{z^2\,dz}{(z^2+1)(z^2+4)} = 2\pi i(B_1 + B_2),$$

or

$$\int_{-R}^{R} \frac{x^2\,dx}{(x^2+1)(x^2+4)} = \frac{\pi}{3} - \int_{C_R} \frac{z^2\,dz}{(z^2+1)(z^2+4)}.$$

If z is a point on C_R, then

$$|z^2+1| \ge ||z|^2 - 1| = R^2 - 1 \quad \text{and} \quad |z^2+4| \ge ||z|^2 - 4| = R^2 - 4.$$

Consequently,

$$\left|\int_{C_R} \frac{z^2\,dz}{(z^2+1)(z^2+4)}\right| \le \frac{\pi R^3}{(R^2-1)(R^2-4)} = \frac{\frac{\pi}{R}}{\left(1-\frac{1}{R^2}\right)\left(1-\frac{4}{R^2}\right)} \to 0 \text{ as } R \to \infty;$$

and we may conclude that

$$\int_{-\infty}^{\infty} \frac{x^2\,dx}{(x^2+1)(x^2+4)} = \frac{\pi}{3}, \quad \text{or} \quad \int_0^{\infty} \frac{x^2\,dx}{(x^2+1)(x^2+4)} = \frac{\pi}{6}.$$

5. The integral $\int_0^{\infty} \frac{x^2\,dx}{(x^2+9)(x^2+4)^2}$ can be evaluated with the aid of the function

$$f(z) = \frac{z^2}{(z^2+9)(z^2+4)^2}$$

and the simple closed contour shown below, where $R > 3$.

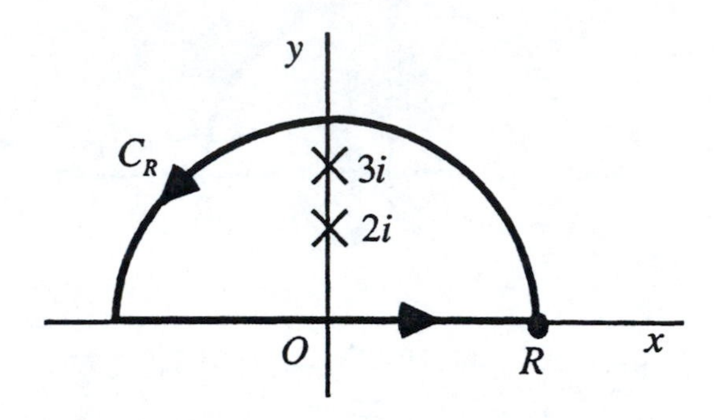

We start by writing

$$\int_{-R}^{R} \frac{x^2\,dx}{(x^2+9)(x^2+4)^2} + \int_{C_R} \frac{z^2\,dz}{(z^2+9)(z^2+4)^2} = 2\pi i(B_1 + B_2),$$

where

$$B_1 = \operatorname*{Res}_{z=3i} \frac{z^2}{(z^2+9)(z^2+4)^2} \quad \text{and} \quad B_2 = \operatorname*{Res}_{z=2i} \frac{z^2}{(z^2+9)(z^2+4)^2}.$$

Now

$$B_1 = \left.\frac{z^2}{(z+3i)(z^2+4)^2}\right]_{z=3i} = -\frac{3}{50i}.$$

To find B_2, we write

$$\frac{z^2}{(z^2+9)(z^2+4)^2} = \frac{\phi(z)}{(z-2i)^2}, \quad \text{where} \quad \phi(z) = \frac{z^2}{(z^2+9)(z+2i)^2}.$$

Then

$$B_2 = \phi'(2i) = \frac{13}{200i}.$$

This tells us that

$$\int_{-R}^{R} \frac{x^2\,dx}{(x^2+9)(x^2+4)^2} = \frac{\pi}{100} - \int_{C_R} \frac{z^2\,dz}{(z^2+9)(z^2+4)^2}.$$

Finally, since

$$\left|\int_{C_R} \frac{z^2\,dz}{(z^2+9)(z^2+4)^2}\right| \le \frac{\pi R^3}{(R^2-9)(R^2-4)^2} \to 0 \text{ as } R \to \infty,$$

we find that

$$\int_{-\infty}^{\infty} \frac{x^2\,dx}{(x^2+9)(x^2+4)^2} = \frac{\pi}{100}, \quad \text{or} \quad \int_0^{\infty} \frac{x^2\,dx}{(x^2+9)(x^2+4)^2} = \frac{\pi}{200}.$$

7. In order to show that

$$\text{P.V.}\int_{-\infty}^{\infty}\frac{x\,dx}{(x^2+1)(x^2+2x+2)}=-\frac{\pi}{5},$$

we introduce the function

$$f(z)=\frac{z}{(z^2+1)(z^2+2z+2)}$$

and the simple closed contour shown below.

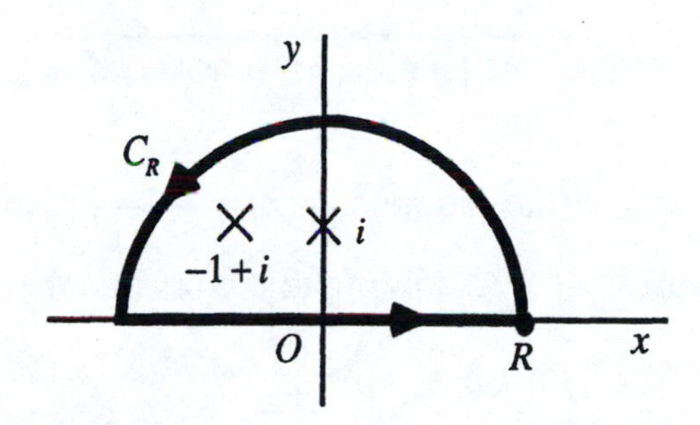

Observe that the singularities of $f(z)$ are at i, $z_0=-1+i$ and their conjugates $-i$, $\bar{z}_0=-1-i$ in the lower half plane. Also, if $R>\sqrt{2}$, we see that

$$\int_{-R}^{R} f(x)\,dx+\int_{C_R} f(z)\,dz=2\pi i(B_0+B_1),$$

where

$$B_0=\operatorname*{Res}_{z=z_0} f(z)=\frac{z}{(z^2+1)(z-\bar{z}_0)}\Bigg]_{z=z_0}=-\frac{1}{10}+\frac{3}{10}i$$

and

$$B_1=\operatorname*{Res}_{z=i} f(z)=\frac{z}{(z+i)(z^2+2z+2)}\Bigg]_{z=i}=\frac{1}{10}-\frac{1}{5}i.$$

Evidently, then,

$$\int_{-R}^{R}\frac{x\,dx}{(x^2+1)(x^2+2x+2)}=-\frac{\pi}{5}-\int_{C_R}\frac{z\,dz}{(z^2+1)(z^2+2z+2)}.$$

Since

$$\left|\int_{C_R}\frac{z\,dz}{(z^2+1)(z^2+2z+2)}\right|=\left|\int_{C_R}\frac{z\,dz}{(z^2+1)(z-z_0)(z-\bar{z}_0)}\right|\le\frac{\pi R^2}{(R^2-1)(R-\sqrt{2})^2}\to 0$$

as $R\to\infty$, this means that

$$\lim_{R\to\infty}\int_{-R}^{R}\frac{x\,dx}{(x^2+1)(x^2+2x+2)}=-\frac{\pi}{5}.$$

This is the desired result.

8. The problem here is to establish the integration formula $\int_0^\infty \frac{dx}{x^3+1} = \frac{2\pi}{3\sqrt{3}}$ using the simple closed contour shown below, where $R > 1$.

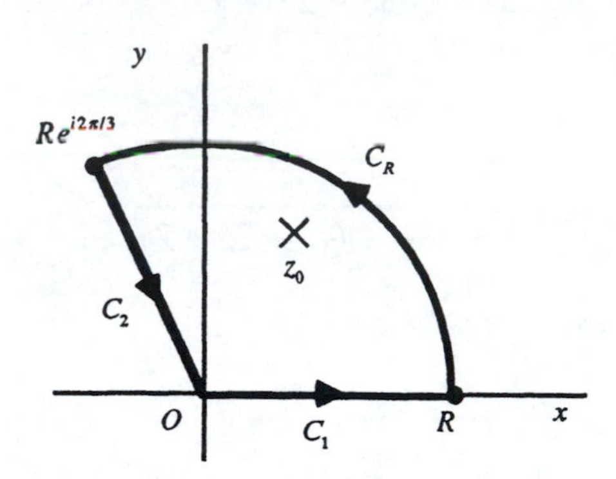

There is only one singularity of the function $f(z) = \frac{1}{z^3+1}$, namely $z_0 = e^{i\pi/3}$, that is interior to the closed contour when $R > 1$. According to the residue theorem,

$$\int_{C_1} \frac{dz}{z^3+1} + \int_{C_R} \frac{dz}{z^3+1} + \int_{C_2} \frac{dz}{z^3+1} = 2\pi i \operatorname*{Res}_{z=z_0} \frac{1}{z^3+1},$$

where the legs of the closed contour are as indicated in the figure. Since C_1 has parametric representation $z = r$ $(0 \le r \le R)$,

$$\int_{C_1} \frac{dz}{z^3+1} = \int_0^R \frac{dr}{r^3+1};$$

and, since $-C_2$ can be represented by $z = re^{i2\pi/3}$ $(0 \le r \le R)$,

$$\int_{C_2} \frac{dz}{z^3+1} = -\int_{-C_2} \frac{dz}{z^3+1} = -\int_0^R \frac{e^{i2\pi/3}dr}{(re^{i2\pi/3})^3+1} = -e^{i2\pi/3}\int_0^R \frac{dr}{r^3+1}.$$

Furthermore,

$$\operatorname*{Res}_{z=z_0} \frac{1}{z^3+1} = \frac{1}{3z_0^2} = \frac{1}{3e^{i2\pi/3}}.$$

Consequently,

$$(1-e^{i2\pi/3})\int_0^R \frac{dr}{r^3+1} = \frac{2\pi i}{3e^{i2\pi/3}} - \int_{C_R} \frac{dz}{z^3+1}.$$

But

$$\left|\int_{C_R} \frac{dz}{z^3+1}\right| \le \frac{1}{R^3-1} \cdot \frac{2\pi R}{3} \to 0 \text{ as } R \to \infty.$$

This gives us the desired result, with the variable of integration r instead of x:

$$\int_0^R \frac{dr}{r^3+1} = \frac{2\pi i}{3(e^{i2\pi/3} - e^{i4\pi/3} \cdot e^{-i6\pi/3})} = \frac{2\pi i}{3(e^{i2\pi/3} - e^{-i2\pi/3})} = \frac{\pi}{3\sin(2\pi/3)} = \frac{2\pi}{3\sqrt{3}}.$$

9. Let m and n be integers, where $0 \le m < n$. The problem here is to derive the integration formula

$$\int_0^\infty \frac{x^{2m}}{x^{2n}+1}\,dx = \frac{\pi}{2n}\csc\left(\frac{2m+1}{2n}\pi\right).$$

(a) The zeros of the polynomial $z^{2n}+1$ occur when $z^{2n}=-1$. Since

$$(-1)^{1/(2n)} = \exp\left[i\frac{(2k+1)\pi}{2n}\right] \qquad (k = 0,1,2,\ldots,2n-1),$$

it is clear that the zeros of $z^{2n}+1$ in the upper half plane are

$$c_k = \exp\left[i\frac{(2k+1)\pi}{2n}\right] \qquad (k = 0,1,2,\ldots,n-1)$$

and that there are none on the real axis.

(b) With the aid of Theorem 2 in Sec. 69, we find that

$$\operatorname*{Res}_{z=c_k}\frac{z^{2m}}{z^{2n}+1} = \frac{c_k^{2m}}{2nc_k^{2n-1}} = \frac{1}{2n}c_k^{2(m-n)+1} \qquad (k = 0,1,2,\ldots,n-1).$$

Putting $\alpha = \dfrac{2m+1}{2n}\pi$, we can write

$$c_k^{2(m-n)+1} = \exp\left[i\frac{(2k+1)\pi(2m-2n+1)}{2n}\right]$$

$$= \exp\left[i\frac{(2k+1)(2m+1)\pi}{2n}\right]\exp[-i(2k+1)\pi] = -e^{i(2k+1)\alpha}.$$

Thus

$$\operatorname*{Res}_{z=c_k}\frac{z^{2m}}{z^{2n}+1} = -\frac{1}{2n}e^{i(2k+1)\alpha} \qquad (k = 0,1,2,\ldots,n-1).$$

In view of the identity (see Exercise 10, Sec. 7)

$$\sum_{k=0}^{n-1} z^k = \frac{1-z^n}{1-z} \qquad (z \ne 1),$$

then,

$$2\pi i\sum_{k=0}^{n-1}\operatorname*{Res}_{z=c_k}\frac{z^{2m}}{z^{2n}+1}=-\frac{\pi i}{n}e^{i\alpha}\sum_{k=0}^{n-1}(e^{i2\alpha})^k=-\frac{\pi i}{n}e^{i\alpha}\frac{1-e^{i2\alpha n}}{1-e^{i2\alpha}}\cdot\frac{e^{-i\alpha}}{e^{-i\alpha}}=-\frac{\pi i}{n}\cdot\frac{e^{i2\alpha n}-1}{e^{i\alpha}-e^{-i\alpha}}$$

$$=-\frac{\pi i}{n}\cdot\frac{e^{i(2m+1)\pi}-1}{e^{i\alpha}-e^{-i\alpha}}=\frac{\pi}{n}\cdot\frac{2i}{e^{i\alpha}-e^{-i\alpha}}=\frac{\pi}{n\sin\alpha}.$$

(c) Consider the path shown below, where $R>1$.

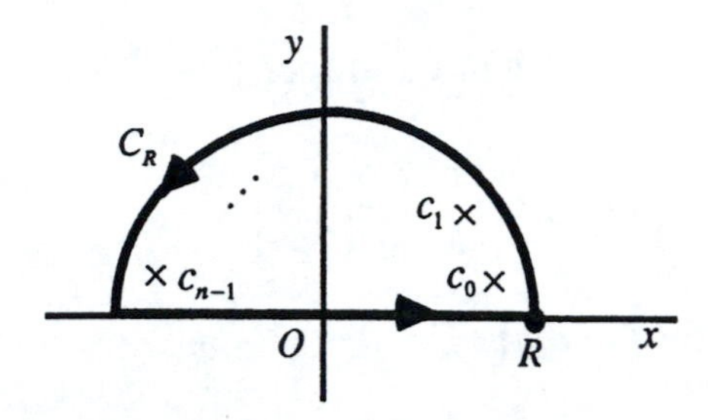

The residue theorem tells us that

$$\int_{-R}^{R}\frac{x^{2m}}{x^{2n}+1}dx+\int_{C_R}\frac{z^{2m}}{z^{2n}+1}dz=2\pi i\sum_{k=0}^{n-1}\operatorname*{Res}_{z=c_k}\frac{z^{2m}}{z^{2n}+1},$$

or

$$\int_{-R}^{R}\frac{x^{2m}}{x^{2n}+1}dx=\frac{\pi}{n\sin\alpha}-\int_{C_R}\frac{z^{2m}}{z^{2n}+1}dz.$$

Observe that if z is a point on C_R, then

$$|z^{2m}|=R^{2m}\quad\text{and}\quad|z^{2n}+1|\geq R^{2n}-1.$$

Consequently,

$$\left|\int_{C_R}\frac{z^{2m}}{z^{2n}+1}dz\right|\leq\frac{R^{2m}}{R^{2n}-1}\pi R\cdot\frac{R^{-2n}}{R^{-2n}}=\pi\frac{\dfrac{1}{R^{2(n-m)-1}}}{1-\dfrac{1}{R^{2n}}}\to 0;$$

and the desired integration formula follows.

10. The problem here is to evaluate the integral

$$\int_0^\infty\frac{dx}{[(x^2-a)^2+1]^2},$$

where a is any real number. We do this by following the steps below.

(a) Let us first find the four zeros of the polynomial

$$q(z) = (z^2 - a)^2 + 1.$$

Solving the equation $q(z) = 0$ for z^2, we obtain $z^2 = a \pm i$. Thus two of the zeros are the square roots of $a + i$, and the other two are the square roots of $a - i$. By Exercise 5, Sec. 9, the two square roots of $a + i$ are the numbers

$$z_0 = \frac{1}{\sqrt{2}}\left(\sqrt{A+a} + i\sqrt{A-a}\right) \quad \text{and} \quad -z_0,$$

where $A = \sqrt{a^2 + 1}$. Since $(\pm \overline{z}_0)^2 = \overline{z_0^2} = \overline{a+i} = a - i$, the two square roots of $a - i$, are evidently

$$\overline{z}_0 \quad \text{and} \quad -\overline{z}_0.$$

The four zeros of $q(z)$ just obtained are located in the plane in the figure below, which tells us that z_0 and $-\overline{z}_0$ lie above the real axis and that the other two zeros lie below it.

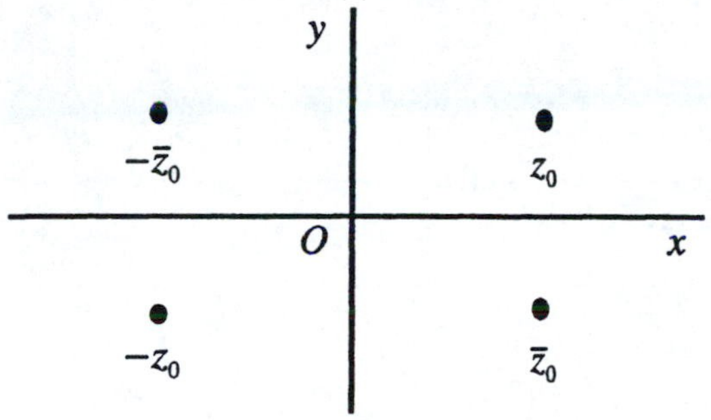

(b) Let $q(z)$ denote the polynomial in part *(a)*; and define the function

$$f(z) = \frac{1}{[q(z)]^2},$$

which becomes the integrand in the integral to be evaluated when $z = x$. The method developed in Exercise 7, Sec. 69, reveals that z_0 is a pole of order 2 of f. To be specific, we note that q is entire and recall from part *(a)* that $q(z_0) = 0$. Furthermore, $q'(z) = 4z(z^2 - a)$ and $z_0^2 = a + i$, as pointed out above in part *(a)*. Consequently, $q'(z_0) = 4z_0(z_0^2 - a) = 4iz_0 \neq 0$. The exercise just mentioned, together with the relations $z_0^2 = a + i$ and $1 + a^2 = A^2$, also enables us to write the residue B_1 of f at z_0:

$$B_1 = -\frac{q''(z_0)}{[q'(z_0)]^3} = -\frac{12z_0^2 - 4a}{(4iz_0)^3} = \frac{3z_0^2 - a}{16iz_0^2 z_0} = \frac{3(a+i) - a}{16i(a+i)z_0} \cdot \frac{a-i}{a-i} = \frac{a - i(2a^2 + 3)}{16A^2 z_0}.$$

As for the point $-\overline{z}_0$, we observe that

$$q'(-\overline{z}) = -\overline{q'(z)} \quad \text{and} \quad q''(-\overline{z}) = \overline{q''(z)}.$$

Since $q(-\overline{z}_0)=0$ and $q'(-\overline{z}_0)=-\overline{q'(z_0)}=4i\overline{z}_0 \neq 0$, the point $-\overline{z}_0$ is also a pole of order 2 of f. Moreover, if B_2 denotes the residue there,

$$B_2 = -\frac{q''(-\overline{z}_0)}{[q'(-\overline{z}_0)]^3} = \frac{\overline{q''(z_0)}}{\overline{[q'(z_0)]^3}} = \overline{\left\{\frac{q''(z_0)}{[q'(z_0)]^3}\right\}} = -\overline{B}_1.$$

Thus

$$B_1 + B_2 = B_1 - \overline{B}_1 = 2i\,\mathrm{Im}\,B_1 = \frac{1}{8A^2 i}\mathrm{Im}\left[\frac{-a+i(2a^2+3)}{z_0}\right].$$

(c) We now integrate $f(z)$ around the simple closed path in the figure below, where $R > |z_0|$ and C_R denotes the semicircular portion of the path. The residue theorem tells us that

$$\int_{-R}^{R} f(x)dx + \int_{C_R} f(z)dz = 2\pi i(B_1 + B_2),$$

or

$$\int_{-R}^{R} \frac{dx}{[(x^2-a)^2+1]^2} = \frac{\pi}{4A^2}\mathrm{Im}\left[\frac{-a+i(2a^2+3)}{z_0}\right] - \int_{C_R}\frac{dz}{[q(z)]^2}.$$

In order to show that

$$\lim_{R\to\infty}\int_{C_R}\frac{dx}{[q(z)]^2} = 0,$$

we start with the observation that the polynomial $q(z)$ can be factored into the form

$$q(z) = (z-z_0)(z+z_0)(z-\overline{z}_0)(z+\overline{z}_0).$$

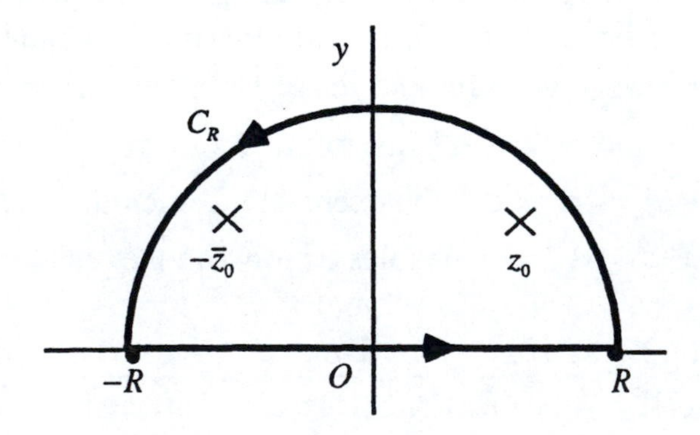

Recall now that $R > |z_0|$. If z is a point on C_R, so that $|z| = R$, then

$$|z \pm z_0| \geq ||z| - |z_0|| = R - |z_0| \quad \text{and} \quad |z \pm \overline{z}_0| \geq ||z| - |\overline{z}_0|| = R - |z_0|.$$

This enables us to see that $|q(z)| \geq (R-|z_0|)^4$ when z is on C_R. Thus

$$\left|\frac{1}{[q(z)]^2}\right| \leq \frac{1}{(R-|z_0|)^8}$$

for such points, and we arrive at the inequality

$$\left|\int_{C_R} \frac{1}{[q(z)]^2}\,dz\right| \leq \frac{\pi R}{(R-|z_0|)^8} = \frac{\dfrac{\pi}{R^7}}{\left(1-\dfrac{|z_0|}{R}\right)^8},$$

which tells us that the value of this integral does, indeed, tend to 0 as R tends to ∞. Consequently,

$$\text{P.V.}\int_{-\infty}^{\infty} \frac{dx}{[(x^2-a)^2+1]^2} = \frac{\pi}{4A^2}\,\text{Im}\left[\frac{-a+i(2a^2+3)}{z_0}\right].$$

But the integrand here is even, and

$$\text{Im}\left[\frac{-a+i(2a^2+3)}{z_0}\right] = \text{Im}\left[\sqrt{2}\,\frac{-a+i(2a^2+3)}{\sqrt{A+a}+i\sqrt{A-a}}\cdot\frac{\sqrt{A+a}-i\sqrt{A-a}}{\sqrt{A+a}-i\sqrt{A-a}}\right].$$

So, the desired result is

$$\int_0^{\infty} \frac{dx}{[(x^2-a)^2+1]^2} = \frac{\pi}{8\sqrt{2}A^3}\left[(2a^2+3)\sqrt{A+a}+a\sqrt{A-a}\right],$$

where $A=\sqrt{a^2+1}$.

SECTION 74

1. The problem here is to evaluate the integral $\displaystyle\int_{-\infty}^{\infty} \frac{\cos x\,dx}{(x^2+a^2)(x^2+b^2)}$, where $a>b>0$. To do this, we introduce the function $f(z)=\dfrac{1}{(z^2+a^2)(z^2+b^2)}$, whose singularities ai and bi lie inside the simple closed contour shown below, where $R>a$. The other singularities are, of course, in the lower half plane.

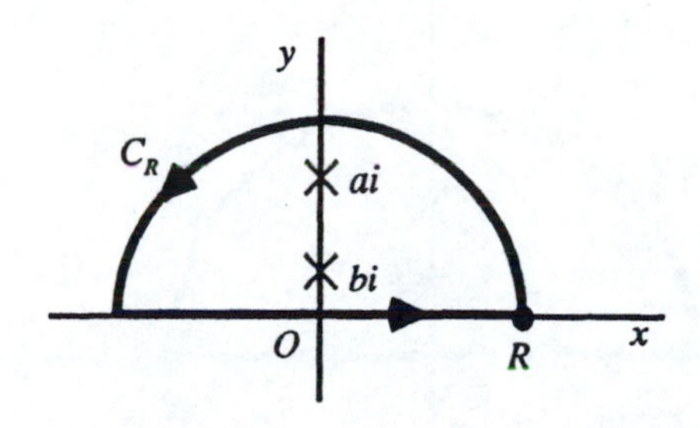

According to the residue theorem,

$$\int_{-R}^{R} \frac{e^{ix}\,dx}{(x^2+a^2)(x^2+b^2)} + \int_{C_R} f(z)e^{iz}dz = 2\pi i(B_1+B_2),$$

where

$$B_1 = \operatorname*{Res}_{z=ai}[f(z)e^{iz}] = \left.\frac{e^{iz}}{(z+ai)(z^2+b^2)}\right]_{z=ai} = \frac{e^{-a}}{2a(b^2-a^2)i}$$

and

$$B_2 = \operatorname*{Res}_{z=bi}[f(z)e^{iz}] = \left.\frac{e^{iz}}{(z^2+a^2)(z+bi)}\right]_{z=bi} = \frac{e^{-b}}{2b(a^2-b^2)i}.$$

That is,

$$\int_{-R}^{R} \frac{e^{ix}\,dx}{(x^2+a^2)(x^2+b^2)} = \frac{\pi}{a^2-b^2}\left(\frac{e^{-b}}{b}-\frac{e^{-a}}{a}\right) - \int_{C_R} f(z)e^{iz}dz,$$

or

$$\int_{-R}^{R} \frac{\cos x\,dx}{(x^2+a^2)(x^2+b^2)} = \frac{\pi}{a^2-b^2}\left(\frac{e^{-b}}{b}-\frac{e^{-a}}{a}\right) - \operatorname{Re}\int_{C_R} f(z)e^{iz}dz.$$

Now, if z is a point on C_R,

$$|f(z)| \le M_R \quad \text{where} \quad M_R = \frac{1}{(R^2-a^2)(R^2-b^2)}$$

and $|e^{iz}| = e^{-y} \le 1$. Hence

$$\left|\operatorname{Re}\int_{C_R} f(z)e^{iz}dz\right| \le \left|\int_{C_R} f(z)e^{iz}dz\right| \le M_R \pi R = \frac{\pi R}{(R^2-a^2)(R^2-b^2)} \to 0 \text{ as } R\to\infty.$$

So it follows that

$$\int_{-\infty}^{\infty} \frac{\cos x\,dx}{(x^2+a^2)(x^2+b^2)} = \frac{\pi}{a^2-b^2}\left(\frac{e^{-b}}{b}-\frac{e^{-a}}{a}\right) \qquad (a>b>0).$$

2. This problem is to evaluate the integral $\int_0^\infty \frac{\cos ax}{x^2+1}dx$, where $a \ge 0$. The function $f(z) = \frac{1}{z^2+1}$ has the singularities $\pm i$; and so we may integrate around the simple closed contour shown below, where $R > 1$.

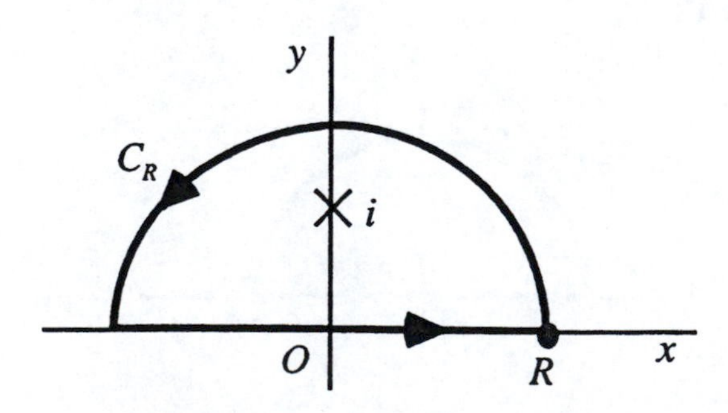

We start with

$$\int_{-R}^{R} \frac{e^{iax}}{x^2+1}dx + \int_{C_R} f(z)e^{iaz}dz = 2\pi iB,$$

where

$$B = \operatorname*{Res}_{z=i}\left[f(z)e^{iaz}\right] = \left.\frac{e^{iaz}}{z+i}\right]_{z=i} = \frac{e^{-a}}{2i}.$$

Hence

$$\int_{-R}^{R} \frac{e^{iax}}{x^2+1}dx = \pi e^{-a} - \int_{C_R} f(z)e^{iaz}dz,$$

or

$$\int_{-R}^{R} \frac{\cos ax}{x^2+1}dx = \pi e^{-a} - \operatorname{Re}\int_{C_R} f(z)e^{iaz}dz,$$

Since

$$|f(z)| \le M_R \quad \text{where} \quad M_R = \frac{1}{R^2-1},$$

we know that

$$\left|\operatorname{Re}\int_{C_R} f(z)e^{iaz}dz\right| \le \left|\int_{C_R} f(z)e^{iaz}dz\right| \le \frac{\pi R}{R^2-1};$$

and so

$$\int_{-\infty}^{\infty} \frac{\cos ax}{x^2+1}dx = \pi e^{-a}.$$

That is,

$$\int_{0}^{\infty} \frac{\cos ax}{x^2+1}dx = \frac{\pi}{2}e^{-a} \qquad (a \ge 0).$$

4. To evaluate the integral $\int_0^\infty \frac{x\sin 2x}{x^2+3}dx$, we first introduce the function

$$f(z) = \frac{z}{z^2+3} = \frac{z}{(z-z_1)(z-\overline{z}_1)},$$

where $z_1 = \sqrt{3}i$. The point z_1 lies above the x axis, and $\overline{z}_1$ lies below it. If we write

$$f(z)e^{i2z} = \frac{\phi(z)}{z-z_1} \quad \text{where} \quad \phi(z) = \frac{z\exp(i2z)}{z-\overline{z}_1},$$

we see that z_1 is a simple pole of the function $f(z)e^{i2z}$ and that the corresponding residue is

$$B_1 = \phi(z_1) = \frac{\sqrt{3}i\exp(-2\sqrt{3})}{2\sqrt{3}i} = \frac{\exp(-2\sqrt{3})}{2}.$$

Now consider the simple closed contour shown in the figure below, where $R > \sqrt{3}$.

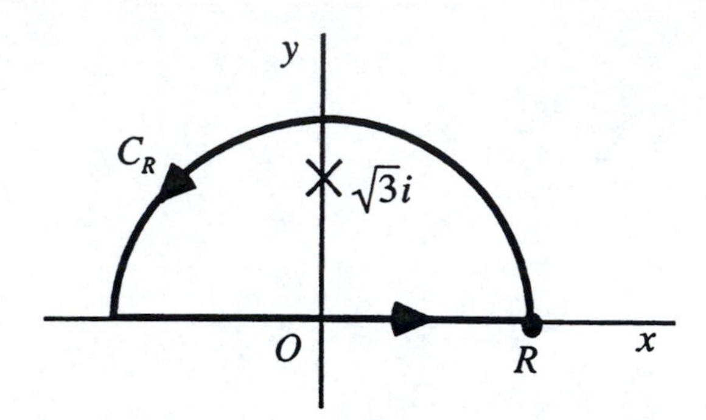

Integrating $f(z)e^{i2z}$ around the closed contour, we have

$$\int_{-R}^{R} \frac{xe^{i2x}}{x^2+3}dx = 2\pi iB_1 - \int_{C_R} f(z)e^{i2z}\,dz.$$

Thus

$$\int_{-R}^{R} \frac{x\sin x}{x^2+3}dx = \text{Im}(2\pi iB_1) - \text{Im}\int_{C_R} f(z)e^{i2z}\,dz.$$

Now, when z is a point on C_R,

$$|f(z)| \le M_R, \quad \text{where} \quad M_R = \frac{R}{R^2-3} \to 0 \text{ as } R \to \infty;$$

and so, by limit (1), Sec. 74,

$$\lim_{R\to\infty} \int_{C_R} f(z)e^{i2z}dz = 0.$$

Consequently, since

$$\left|\text{Im}\int_{C_R} f(z)e^{i2z}\,dz\right| \le \left|\int_{C_R} f(z)e^{i2z}\,dz\right|,$$

we arrive at the result

$$\int_{-\infty}^{\infty} \frac{x\sin x}{x^2+3}dx = \pi\exp(-2\sqrt{3}), \quad \text{or} \quad \int_{0}^{\infty} \frac{x\sin x}{x^2+3}dx = \frac{\pi}{2}\exp(-2\sqrt{3}).$$

6. The integral to be evaluated is $\int_{-\infty}^{\infty} \frac{x^3 \sin ax}{x^4+4} dx$, where $a > 0$. We define the function $f(z) = \frac{z^3}{z^4+4}$; and, by computing the fourth roots of -4, we find that the singularities

$$z_1 = \sqrt{2}e^{i\pi/4} = 1+i \quad \text{and} \quad z_2 = \sqrt{2}e^{i3\pi/4} = \sqrt{2}e^{i\pi/4}e^{i\pi/2} = (1+i)i = -1+i$$

both lie inside the simple closed contour shown below, where $R > \sqrt{2}$. The other two singularities lie below the real axis.

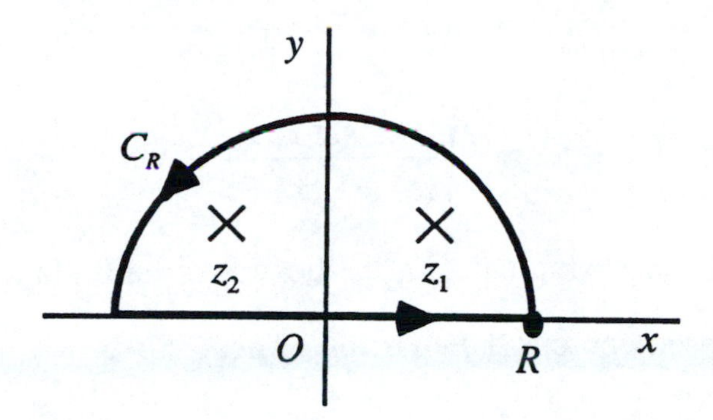

The residue theorem and the method of Theorem 2 in Sec. 69 for finding residues at simple poles tell us that

$$\int_{-R}^{R} \frac{x^3 e^{iax}}{x^4+4} dx + \int_{C_R} f(z)e^{iaz} dz = 2\pi i(B_1 + B_2),$$

where

$$B_1 = \operatorname*{Res}_{z=z_1} \frac{z^3 e^{iaz}}{z^4+4} = \frac{z_1^3 e^{iaz_1}}{4z_1^3} = \frac{e^{iaz_1}}{4} = \frac{e^{ia(1+i)}}{4} = \frac{e^{-a}e^{ia}}{4}$$

and

$$B_2 = \operatorname*{Res}_{z=z_2} \frac{z^3 e^{iaz}}{z^4+4} = \frac{z_2^3 e^{iaz_2}}{4z_2^3} = \frac{e^{iaz_2}}{4} = \frac{e^{ia(-1+i)}}{4} = \frac{e^{-a}e^{-ia}}{4}.$$

Since

$$2\pi i(B_1 + B_2) = \pi i e^{-a}\left(\frac{e^{ia} + e^{-ia}}{2}\right) = i\pi e^{-a} \cos a,$$

we are now able to write

$$\int_{-R}^{R} \frac{x^3 \sin ax}{x^4+4} dx = \pi e^{-a} \cos a - \operatorname{Im} \int_{C_R} f(z)e^{iaz} dz.$$

Furthermore, if z is a point on C_R, then

$$|f(z)| \le M_R \quad \text{where} \quad M_R = \frac{R^3}{R^4 - 4} \to 0 \text{ as } R \to \infty;$$

and this means that

$$\left| \operatorname{Im} \int_{C_R} f(z) e^{iaz} dz \right| \le \left| \int_{C_R} f(z) e^{iaz} dz \right| \to 0 \text{ as } R \to \infty,$$

according to limit (1), Sec. 74. Finally, then,

$$\int_{-\infty}^{\infty} \frac{x^3 \sin ax}{x^4 + 4} dx = \pi e^{-a} \cos a \qquad (a > 0).$$

8. In order to evaluate the integral $\int_0^{\infty} \frac{x^3 \sin x \, dx}{(x^2+1)(x^2+9)}$, we introduce here the function $f(z) = \frac{z^3}{(z^2+1)(z^2+9)}$. Its singularities in the upper half plane are i and $3i$, and we consider the simple closed contour shown below, where $R > 3$.

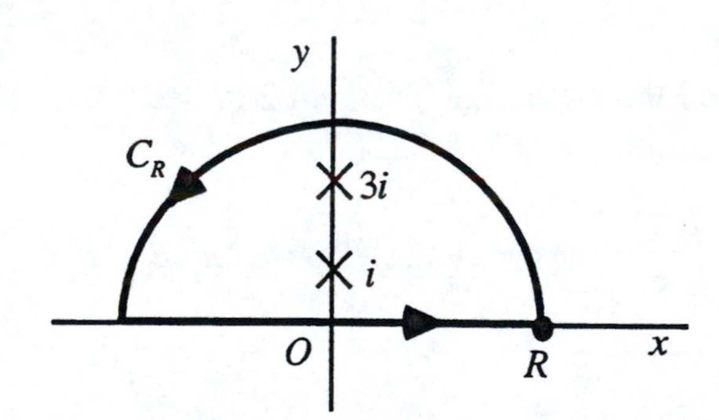

Since

$$\operatorname*{Res}_{z=i}\left[f(z)e^{iz}\right] = \frac{z^3 e^{iz}}{(z+i)(z^2+9)}\Bigg]_{z=i} = -\frac{1}{16e}$$

and

$$\operatorname*{Res}_{z=3i}\left[f(z)e^{iz}\right] = \frac{z^3 e^{iz}}{(z^2+1)(z+3i)}\Bigg]_{z=3i} = \frac{9}{16e^3},$$

the residue theorem tells us that

$$\int_{-R}^{R} \frac{x^3 e^{ix} \, dx}{(x^2+1)(x^2+9)} + \int_{C_R} f(z) e^{iz} dx = 2\pi i \left(-\frac{1}{16e} + \frac{9}{16e^3} \right),$$

or

$$\int_{-R}^{R} \frac{x^3 \sin x \, dx}{(x^2+1)(x^2+9)} = \frac{\pi}{8e}\left(\frac{9}{e^2} - 1\right) - \operatorname{Im} \int_{C_R} f(z) e^{iz} dz.$$

Now if z is a point on C_R, then

$$|f(z)| \leq M_R \quad \text{where} \quad M_R = \frac{R}{(R^2-1)(R^2-9)} \text{ as } R \to \infty.$$

So, in view of limit (1), Sec. 74,

$$\left|\text{Im} \int_{C_R} f(z)e^{iz}dz\right| \leq \left|\int_{C_R} f(z)e^{iz}dz\right| \to 0 \text{ as } R \to \infty;$$

and this means that

$$\int_{-\infty}^{\infty} \frac{x^3 \sin x \, dx}{(x^2+1)(x^2+9)} = \frac{\pi}{8e}\left(\frac{9}{e^2}-1\right), \quad \text{or} \quad \int_{0}^{\infty} \frac{x^3 \sin x \, dx}{(x^2+1)(x^2+9)} = \frac{\pi}{16e}\left(\frac{9}{e^2}-1\right).$$

9. The Cauchy principal value of the integral $\int_{-\infty}^{\infty} \frac{\sin x \, dx}{x^2+4x+5}$ can be found with the aid of the function $f(z) = \frac{1}{z^2+4z+5}$ and the simple closed contour shown below, where $R > \sqrt{5}$. Using the quadratic formula to solve the equation $z^2+4z+5=0$, we find that f has singularities at the points $z_1 = -2+i$ and $\overline{z}_1 = -2-i$. Thus $f(z) = \frac{1}{(z-z_1)(z-\overline{z}_1)}$, where z_1 is interior to the closed contour and $\overline{z}_1$ is below the real axis.

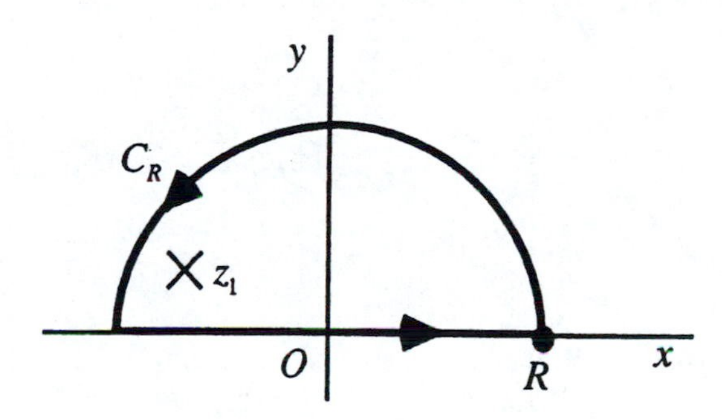

The residue theorem tells us that

$$\int_{-R}^{R} \frac{e^{ix} \, dx}{x^2+4x+5} + \int_{C_R} f(z)e^{iz}dz = 2\pi i B,$$

where

$$B = \underset{z=z_1}{\text{Res}}\left[\frac{e^{iz}}{(z-z_1)(z-\overline{z}_1)}\right] = \frac{e^{iz_1}}{(z_1-\overline{z}_1)};$$

and so

$$\int_{-R}^{R} \frac{\sin x \, dx}{x^2+4x+5} = \text{Im}\left[\frac{2\pi i e^{iz_1}}{(z_1-\overline{z}_1)}\right] - \text{Im} \int_{C_R} f(z)e^{iz}dz,$$

or

$$\int_{-R}^{R} \frac{\sin x\,dx}{x^2+4x+5} = -\frac{\pi}{e}\sin 2 - \operatorname{Im}\int_{C_R} f(z)e^{iz}dz.$$

Now, if z is a point on C_R, then $|e^{iz}| = e^{-y} \le 1$ and

$$|f(z)| \le M_R \quad \text{where} \quad M_R = \frac{1}{(R-|z_1|)(R-|\overline{z}_1|)} = \frac{1}{(R-\sqrt{5})^2}.$$

Hence

$$\left|\operatorname{Im}\int_{C_R} f(z)e^{iz}dz\right| \le \left|\int_{C_R} f(z)e^{iz}dz\right| \le M_R \pi R = \frac{\pi R}{(R-\sqrt{5})^2} \to 0 \text{ as } R \to \infty,$$

and we may conclude that

$$\text{P.V.}\int_{-\infty}^{\infty} \frac{\sin x\,dx}{x^2+4x+5} = -\frac{\pi}{e}\sin 2.$$

10. To find the Cauchy principal value of the improper integral $\int_{-\infty}^{\infty} \frac{(x+1)\cos x}{x^2+4x+5}dx$, we shall use the function $f(z) = \frac{z+1}{z^2+4z+5} = \frac{z+1}{(z-z_1)(z-\overline{z}_1)}$, where $z_1 = -2+i$, and $\overline{z}_1 = -2-1$, and the same simple closed contour as in Exercise 9. In this case,

$$\int_{-R}^{R} \frac{(x+1)e^{ix}\,dx}{x^2+4x+5} + \int_{C_R} f(z)e^{iz}dz = 2\pi iB,$$

where

$$B = \operatorname*{Res}_{z=z_1}\left[\frac{(z+1)e^{iz}}{(z-z_1)(z-\overline{z}_1)}\right] = \frac{(z_1+1)e^{iz_1}}{(z-\overline{z}_1)} = \frac{(-1+i)e^{-2i}}{2ei}.$$

Thus

$$\int_{-R}^{R} \frac{(x+1)\cos x}{x^2+4x+5}dx = \operatorname{Re}(2\pi iB) - \int_{C_R} f(z)e^{iz},$$

or

$$\int_{-R}^{R} \frac{(x+1)\cos x}{x^2+4x+5}dx = \frac{\pi}{e}(\sin 2 - \cos 2) - \int_{C_R} f(z)e^{iz}dz.$$

Finally, we observe that if z is a point on C_R, then

$$|f(z)| \le M_R \quad \text{where} \quad M_R = \frac{R+1}{(R-|z_1|)(R-|\overline{z}_1|)} = \frac{R+1}{(R-\sqrt{5})^2} \to 0 \text{ as } R \to \infty.$$

Limit (1), Sec. 74, then tells us that

$$\left|\operatorname{Re}\int_{C_R} f(z)e^{iz}dz\right| \le \left|\int_{C_R} f(z)e^{iz}dz\right| \to 0 \text{ as } R\to\infty,$$

and so

$$\text{P.V.}\int_{-\infty}^{\infty}\frac{(x+1)\cos x}{x^2+4x+5}dx = \frac{\pi}{e}(\sin 2 - \cos 2).$$

12. *(a)* Since the function $f(z)=\exp(iz^2)$ is entire, the Cauchy-Goursat theorem tells us that its integral around the positively oriented boundary of the sector $0\le r\le R,\ 0\le\theta\le\pi/4$ has value zero. The closed path is shown below.

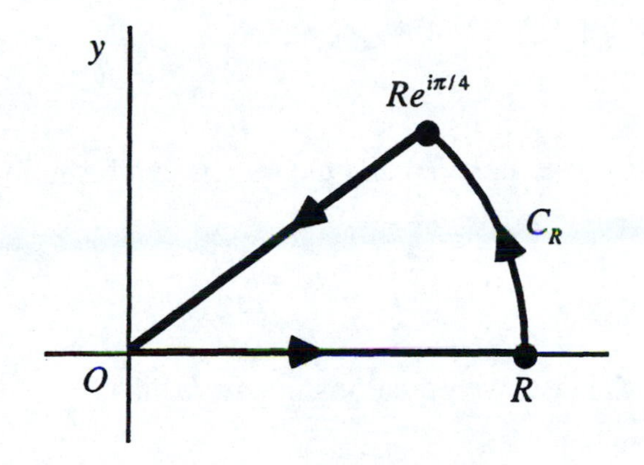

A parametric representation of the horizontal line segment from the origin to the point R is $z=x$ $(0\le x\le R)$, and a representation for the segment from the origin to the point $Re^{i\pi/4}$ is $z=re^{i\pi/4}$ $(0\le r\le R)$. Thus

$$\int_0^R e^{ix^2}dx + \int_{C_R} e^{iz^2}dz - e^{i\pi/4}\int_0^R e^{-r^2}dr = 0,$$

or

$$\int_0^R e^{ix^2}dx = e^{i\pi/4}\int_0^R e^{-r^2}dr - \int_{C_R} e^{iz^2}dz.$$

By equating real parts and then imaginary parts on each side of this last equation, we see that

$$\int_0^R \cos(x^2)\,dx = \frac{1}{\sqrt{2}}\int_0^R e^{-r^2}dr - \operatorname{Re}\int_{C_R} e^{iz^2}dz$$

and

$$\int_0^R \sin(x^2)\,dx = \frac{1}{\sqrt{2}}\int_0^R e^{-r^2}dr - \operatorname{Im}\int_{C_R} e^{iz^2}dz.$$

(b) A parametric representation for the arc C_R is $z = Re^{i\theta}$ $(0 \le \theta \le \pi/4)$. Hence

$$\int_{C_R} e^{iz^2}\,dz = \int_0^{\pi/4} e^{iR^2 e^{i2\theta}} Rie^{i\theta}\,d\theta = iR\int_0^{\pi/4} e^{-R^2\sin 2\theta} e^{iR^2\cos 2\theta} e^{i\theta}\,d\theta.$$

Since $\left|e^{iR^2\cos 2\theta}\right| = 1$ and $\left|e^{i\theta}\right| = 1$, it follows that

$$\left|\int_{C_R} e^{iz^2}\,dz\right| \le R\int_0^{\pi/4} e^{-R^2\sin 2\theta}\,d\theta.$$

Then, by making the substitution $\phi = 2\theta$ in this last integral and referring to the form (3), Sec. 74, of Jordan's inequality, we find that

$$\left|\int_{C_R} e^{iz^2}\,dz\right| \le \frac{R}{2}\int_0^{\pi/2} e^{-R^2\sin\phi}\,d\phi \le \frac{R}{2}\cdot\frac{\pi}{2R^2} = \frac{\pi}{4R} \to 0 \text{ as } R \to \infty.$$

(c) In view of the result in part *(b)* and the integration formula

$$\int_0^{\infty} e^{-x^2}\,dx = \frac{\sqrt{\pi}}{2},$$

it follows from the last two equations in part *(a)* that

$$\int_0^{\infty} \cos(x^2)\,dx = \frac{1}{2}\sqrt{\frac{\pi}{2}} \quad \text{and} \quad \int_0^{\infty} \sin(x^2)\,dx = \frac{1}{2}\sqrt{\frac{\pi}{2}}.$$

SECTION 77

1. The main problem here is to derive the integration formula

$$\int_0^{\infty} \frac{\cos(ax) - \cos(bx)}{x^2}\,dx = \frac{\pi}{2}(b-a) \qquad (a \ge 0, b \ge 0),$$

using the indented contour shown below.

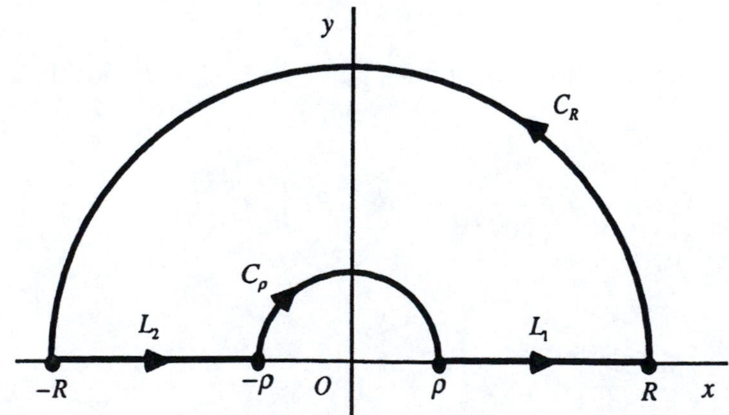

Applying the Cauchy-Goursat theorem to the function

$$f(z) = \frac{e^{iaz} - e^{ibz}}{z^2},$$

we have

$$\int_{L_1} f(z)\,dz + \int_{C_R} f(z)\,dz + \int_{L_2} f(z)\,dz + \int_{C_\rho} f(z)\,dz = 0,$$

or

$$\int_{L_1} f(z)\,dz + \int_{L_2} f(z)\,dz = -\int_{C_\rho} f(z)\,dz - \int_{C_R} f(z)\,dz.$$

Since L_1 and $-L_2$ have parametric representations

$$L_1 : z = re^{i0} = r\ (\rho \le r \le R) \quad \text{and} \quad -L_2 : z = re^{i\pi} = -r\ (\rho \le r \le R),$$

we can see that

$$\int_{L_1} f(z)\,dz + \int_{L_2} f(z)\,dz = \int_{L_1} f(z)\,dz - \int_{-L_2} f(z)\,dz = \int_\rho^R \frac{e^{iar} - e^{ibr}}{r^2}\,dr + \int_\rho^R \frac{e^{-iar} - e^{-ibr}}{r^2}\,dr$$

$$= \int_\rho^R \frac{(e^{iar} + e^{-iar}) - (e^{ibr} + e^{-ibr})}{r^2}\,dr = 2\int_\rho^R \frac{\cos(ar) - \cos(br)}{r^2}\,dr.$$

Thus

$$2\int_\rho^R \frac{\cos(ar) - \cos(br)}{r^2}\,dr = -\int_{C_\rho} f(z)\,dz - \int_{C_R} f(z)\,dz.$$

In order to find the limit of the first integral on the right here as $\rho \to 0$, we write

$$f(z) = \frac{1}{z^2}\left[\left(1 + \frac{iaz}{1!} + \frac{(iaz)^2}{2!} + \frac{(iaz)^3}{3!} + \cdots\right) - \left(1 + \frac{ibz}{1!} + \frac{(ibz)^2}{2!} + \frac{(ibz)^3}{3!} + \cdots\right)\right]$$

$$= \frac{i(a-b)}{z} + \cdots \quad (0 < |z| < \infty).$$

From this we see that $z = 0$ is a simple pole of $f(z)$, with residue $B_0 = i(a-b)$. Thus

$$\lim_{\rho \to 0} \int_{C_\rho} f(z)\,dz = -B_0\pi i = -i(a-b)\pi i = \pi(a-b).$$

As for the limit of the value of the second integral as $R \to \infty$, we note that if z is a point on C_R, then

$$f(z) \le \frac{|e^{iaz}|+|e^{ibz}|}{|z|^2} = \frac{e^{-ay}+e^{-by}}{R^2} \le \frac{1+1}{R^2} = \frac{2}{R^2}.$$

Consequently,

$$\left| \int_{C_R} f(z)\,dz \right| \le \frac{2}{R^2}\pi R = \frac{2\pi}{R} \to 0 \text{ as } R \to \infty.$$

It is now clear that letting $\rho \to 0$ and $R \to \infty$ yields

$$2\int_0^\infty \frac{\cos(ar)-\cos(br)}{r^2}\,dr = \pi(b-a).$$

This is the desired integration formula, with the variable of integration r instead of x. Observe that when $a = 0$ and $b = 2$, that result becomes

$$\int_0^\infty \frac{1-\cos(2x)}{x^2}\,dx = \pi.$$

But $\cos(2x) = 1 - 2\sin^2 x$, and we arrive at

$$\int_0^\infty \frac{\sin^2 x}{x^2}\,dx = \frac{\pi}{2}.$$

2. Let us derive the integration formula

$$\int_0^\infty \frac{x^a}{(x^2+1)^2}\,dx = \frac{(1-a)\pi}{4\cos(a\pi/2)} \qquad (-1 < a < 3),$$

where $x^a = \exp(a \ln x)$ when $x > 0$. We shall integrate the function

$$f(z) = \frac{z^a}{(z^2+1)^2} = \frac{\exp(a \log z)}{(z^2+1)^2} \qquad \left(|z| > 0, -\frac{\pi}{2} < \arg z < \frac{3\pi}{2}\right),$$

whose branch cut is the origin and the negative imaginary axis, around the simple closed path shown below.

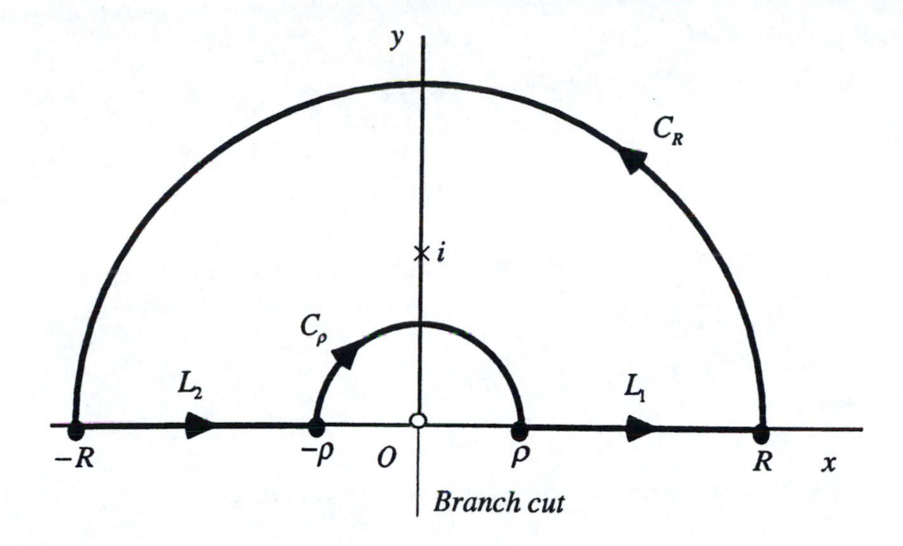

By Cauchy's residue theorem,

$$\int_{L_1} f(z)\,dz + \int_{C_R} f(z)\,dz + \int_{L_2} f(z)\,dz + \int_{C_\rho} f(z)\,dz = 2\pi i \operatorname*{Res}_{z=i} f(z).$$

That is,

$$\int_{L_1} f(z)\,dz + \int_{L_2} f(z)\,dz = 2\pi i \operatorname*{Res}_{z=i} f(z) - \int_{C_\rho} f(z)\,dz - \int_{C_R} f(z)\,dz.$$

Since

$$L_1: z = re^{i0} = r\ (\rho \le r \le R) \quad \text{and} \quad -L_2: z = re^{i\pi} = -r\ (\rho \le r \le R),$$

the left-hand side of this last equation can be written

$$\int_{L_1} f(z)\,dz - \int_{-L_2} f(z)\,dz = \int_\rho^R \frac{e^{a(\ln r + i0)}}{(r^2+1)^2}\,dr - \int_\rho^R \frac{e^{a(\ln r + i\pi)}}{(r^2+1)^2} e^{i\pi}\,dr$$

$$= \int_\rho^R \frac{r^a}{(r^2+1)^2}\,dr + e^{ia\pi} \int_\rho^R \frac{r^a}{(r^2+1)^2}\,dr = (1+e^{ia\pi}) \int_\rho^R \frac{r^a}{(r^2+1)^2}\,dr.$$

Also,

$$\operatorname*{Res}_{z=i} f(z) = \phi'(i) \quad \text{where} \quad \phi(z) = \frac{z^a}{(z+i)^2},$$

the point $z = i$ being a pole of order 2 of the function $f(z)$. Straightforward differentiation reveals that

$$\phi'(z) = e^{(a-1)\log z}\left[\frac{a(z+i) - 2z}{(z+i)^3}\right],$$

and from this it follows that

$$\operatorname*{Res}_{z=i} f(z) = -ie^{ia\pi/2}\left(\frac{1-a}{4}\right).$$

We now have

$$(1+e^{ia\pi})\int_{\rho}^{R}\frac{r^a}{(r^2+1)^2}\,dr = \frac{\pi(1-a)}{2}e^{ia\pi/2} - \int_{C_\rho} f(z)dz - \int_{C_R} f(z)dz.$$

Once we show that

$$\lim_{\rho\to 0}\int_{C_\rho} f(z)\,dz = 0 \quad \text{and} \quad \lim_{R\to\infty}\int_{C_R} f(z)\,dz = 0,$$

we arrive at the desired result:

$$\int_0^\infty \frac{r^a}{(r^2+1)^2}\,dr = \frac{\pi(1-a)}{2}\cdot\frac{e^{ia\pi/2}}{1+e^{ia\pi}}\cdot\frac{e^{-ia\pi/2}}{e^{-ia\pi/2}} = \frac{\pi(1-a)}{4}\cdot\frac{2}{e^{ia\pi/2}+e^{-ia\pi/2}} = \frac{(1-a)\pi}{4\cos(a\pi/2)}.$$

The first of the above limits is shown by writing

$$\left|\int_{C_\rho} f(z)\,dz\right| \le \frac{\rho^a}{(1-\rho^2)^2}\pi\rho = \frac{\pi\rho^{a+1}}{(1-\rho^2)^2}$$

and noting that the last term tends to 0 as $\rho\to 0$ since $a+1>0$. As for the second limit,

$$\left|\int_{C_R} f(z)\,dz\right| \le \frac{R^a}{(R^2-1)^2}\pi R = \frac{\pi R^{a+1}}{(R^2-1)^2}\cdot\frac{\frac{1}{R^4}}{\frac{1}{R^4}} = \frac{\pi\frac{1}{R^{3-a}}}{\left(1-\frac{1}{R^2}\right)^2};$$

and the last term here tends to 0 as $R\to\infty$ since $3-a>0$.

3. The problem here is to derive the integration formulas

$$I_1 = \int_0^\infty \frac{\sqrt[3]{x}\ln x}{x^2+1}\,dx = \frac{\pi^2}{6} \quad \text{and} \quad I_2 = \int_0^\infty \frac{\sqrt[3]{x}}{x^2+1}\,dx = \frac{\pi}{\sqrt{3}}$$

by integrating the function

$$f(z) = \frac{z^{1/3}\log z}{z^2+1} = \frac{e^{(1/3)\log z}\log z}{z^2+1} \qquad \left(|z|>0, -\frac{\pi}{2}<\arg z<\frac{3\pi}{2}\right),$$

around the contour shown in Exercise 2. As was the case in that exercise,

$$\int_{L_1} f(z)dz + \int_{L_2} f(z)dz = 2\pi i \operatorname*{Res}_{z=i} f(z) - \int_{C_\rho} f(z)dz - \int_{C_R} f(z)dz.$$

Since

$$f(z) = \frac{\phi(z)}{z-i} \quad \text{where} \quad \phi(z) = \frac{e^{(1/3)\log z}\log z}{z+i},$$

the point $z = i$ is a simple pole of $f(z)$, with residue

$$\operatorname*{Res}_{z=i} f(z) = \phi(i) = \frac{\pi}{4}e^{i\pi/6}.$$

The parametric representations

$$L_1 : z = re^{i0} = r \ (\rho \le r \le R) \quad \text{and} \quad -L_2 : z = re^{i\pi} = -r \ (\rho \le r \le R)$$

can be used to write

$$\int_{L_1} f(z)dz = \int_\rho^R \frac{\sqrt[3]{r}\ln r}{r^2+1}dr \quad \text{and} \quad \int_{L_2} f(z)dz = e^{i\pi/3}\int_\rho^R \frac{\sqrt[3]{r}\ln r + i\pi\sqrt[3]{r}}{r^2+1}dr.$$

Thus

$$\int_\rho^R \frac{\sqrt[3]{r}\ln r}{r^2+1}dr + e^{i\pi/3}\int_\rho^R \frac{\sqrt[3]{r}\ln r + i\pi\sqrt[3]{r}}{r^2+1}dr = \frac{\pi^2}{2}ie^{i\pi/6} - \int_{C_\rho} f(z)dz - \int_{C_R} f(z)dz.$$

By equating real parts on each side of this equation, we have

$$\int_\rho^R \frac{\sqrt[3]{r}\ln r}{r^2+1}dr + \cos(\pi/3)\int_\rho^R \frac{\sqrt[3]{r}\ln r}{r^2+1}dr - \pi\sin(\pi/3)\int_\rho^R \frac{\sqrt[3]{r}}{r^2+1}dr = -\frac{\pi^2}{2}\sin(\pi/6)$$
$$- \operatorname{Re}\int_{C_\rho} f(z)dz - \operatorname{Re}\int_{C_R} f(z)dz;$$

and equating imaginary parts yields

$$\sin(\pi/3)\int_\rho^R \frac{\sqrt[3]{r}\ln r}{r^2+1}dr + \pi\cos(\pi/3)\int_\rho^R \frac{\sqrt[3]{r}}{r^2+1}dr = \frac{\pi^2}{2}\cos(\pi/6)$$
$$- \operatorname{Im}\int_{C_\rho} f(z)dz - \operatorname{Im}\int_{C_R} f(z)dz.$$

Now $\sin(\pi/3) = \frac{\sqrt{3}}{2}$, $\cos(\pi/3) = \frac{1}{2}$, $\sin(\pi/6) = \frac{1}{2}$, $\cos(\pi/6) = \frac{\sqrt{3}}{2}$ and it is routine to show that

$$\lim_{\rho \to 0} \int_{C_\rho} f(z)dz = 0 \quad \text{and} \quad \lim_{R\to\infty} \int_{C_R} f(z)dz = 0.$$

Thus

$$\frac{3}{2}\int_0^\infty \frac{\sqrt[3]{r}\ln r}{r^2+1}dr - \frac{\pi\sqrt{3}}{2}\int_0^\infty \frac{\sqrt[3]{r}}{r^2+1}dr = -\frac{\pi^2}{4},$$

$$\frac{\sqrt{3}}{2}\int_0^\infty \frac{\sqrt[3]{r}\ln r}{r^2+1}dr + \frac{\pi}{2}\int_0^\infty \frac{\sqrt[3]{r}}{r^2+1}dr = \frac{\pi^2\sqrt{3}}{4}.$$

That is,

$$\frac{3}{2}I_1 - \frac{\pi\sqrt{3}}{2}I_2 = -\frac{\pi^2}{4},$$

$$\frac{\sqrt{3}}{2}I_1 + \frac{\pi}{2}I_2 = \frac{\pi^2\sqrt{3}}{4}.$$

Solving these simultaneous equations for I_1 and I_2, we arrive at the desired integration formulas.

4. Let us use the function

$$f(z) = \frac{(\log z)^2}{z^2+1} \qquad \left(|z|>0, -\frac{\pi}{2} < \arg z < \frac{3\pi}{2}\right)$$

and the contour in Exercise 2 to show that

$$\int_0^\infty \frac{(\ln x)^2}{x^2+1}dx = \frac{\pi^3}{8} \quad \text{and} \quad \int_0^\infty \frac{\ln x}{x^2+1}dx = 0.$$

Integrating $f(z)$ around the closed path shown in Exercise 2, we have

$$\int_{L_1} f(z)\,dz + \int_{L_2} f(z)\,dz = 2\pi i \operatorname*{Res}_{z=i} f(z) - \int_{C_\rho} f(z)\,dz - \int_{C_R} f(z)\,dz.$$

Since

$$f(z) = \frac{\phi(z)}{z-i} \quad \text{where} \quad \phi(z) = \frac{(\log z)^2}{z+i},$$

the point $z=i$ is a simple pole of $f(z)$ and the residue is

$$\operatorname*{Res}_{z=i} f(z) = \phi(i) = \frac{(\log i)^2}{2i} = \frac{(\ln 1 + i\pi/2)^2}{2i} = -\frac{\pi^2}{8i}.$$

Also, the parametric representations

$$L_1 : z = re^{i0} = r \ (\rho \le r \le R) \quad \text{and} \quad -L_2 : z = re^{i\pi} = -r \ (\rho \le r \le R)$$

enable us to write

$$\int_{L_1} f(z)dz = \int_\rho^R \frac{(\ln r)^2}{r^2+1}dr \quad \text{and} \quad \int_{L_2} f(z)dz = \int_\rho^R \frac{(\ln r + i\pi)^2}{r^2+1}dr.$$

Since

$$\int_{L_1} f(z)dz + \int_{L_2} f(z)dz = 2\int_\rho^R \frac{(\ln r)^2}{r^2+1}dr - \pi^2\int_\rho^R \frac{dr}{r^2+1} + 2\pi i\int_\rho^R \frac{\ln r}{r^2+1}dr,$$

then,

$$2\int_\rho^R \frac{(\ln r)^2}{r^2+1}dr - \pi^2\int_\rho^R \frac{dr}{r^2+1} + 2\pi i\int_\rho^R \frac{\ln r}{r^2+1}dr = -\frac{\pi^3}{4} - \int_{C_\rho} f(z)dz - \int_{C_R} f(z)dz.$$

Equating real parts on each side of this equation, we have

$$2\int_\rho^R \frac{(\ln r)^2}{r^2+1}dr - \pi^2\int_\rho^R \frac{dr}{r^2+1} = -\frac{\pi^3}{4} - \text{Re}\int_{C_\rho} f(z)dz - \text{Re}\int_{C_R} f(z)dz;$$

and equating imaginary parts yields

$$2\pi\int_\rho^R \frac{\ln r}{r^2+1}dr = \text{Im}\int_{C_\rho} f(z)dz - \text{Im}\int_{C_R} f(z)dz.$$

It is straightforward to show that

$$\lim_{\rho\to 0}\int_{C_\rho} f(z)dz = 0 \quad \text{and} \quad \lim_{R\to\infty}\int_{C_R} f(z)dz = 0.$$

Hence

$$2\int_0^\infty \frac{(\ln r)^2}{r^2+1}dr - \pi^2\int_0^\infty \frac{dr}{r^2+1} = -\frac{\pi^3}{4}$$

and

$$2\pi\int_0^\infty \frac{\ln r}{r^2+1}dr = 0.$$

Finally, inasmuch as (see Exercise 1, Sec. 72),

$$\int_0^\infty \frac{dr}{r^2+1} = \frac{\pi}{2},$$

we arrive at the desired integration formulas.

5. Here we evaluate the integral $\int_0^\infty \frac{\sqrt[3]{x}}{(x+a)(x+b)}dx$, where $a>b>0$. We consider the function

$$f(z)=\frac{z^{1/3}}{(z+a)(z+b)}=\frac{\exp\left(\frac{1}{3}\log z\right)}{(z+a)(z+b)} \qquad (|z|>0, 0<\arg z<2\pi)$$

and the simple closed contour shown below, which is similar to the one used in Sec. 77. The numbers ρ and R are small and large enough, respectively, so that the points $z=-a$ and $z=-b$ are between the circles.

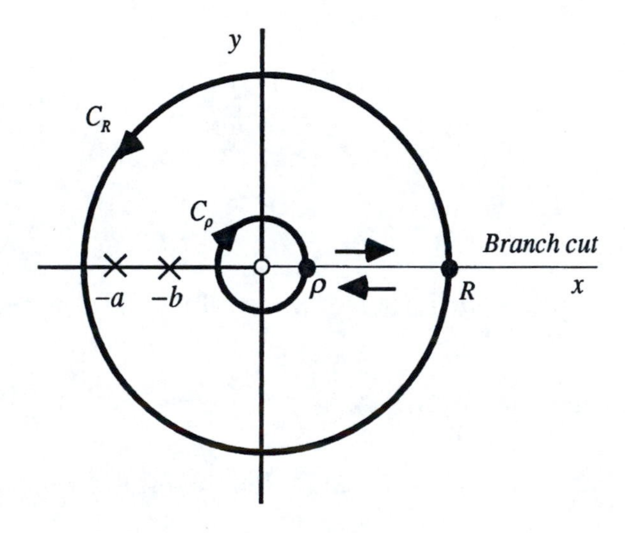

A parametric representation for the upper edge of the branch cut from ρ to R is $z=re^{i0}$ $(\rho\le r\le R)$, and so the value of the integral of *f* along that edge is

$$\int_\rho^R \frac{\exp\left[\frac{1}{3}(\ln r+i0)\right]}{(r+a)(r+b)}dr=\int_\rho^R\frac{\sqrt[3]{r}}{(r+a)(r+b)}dr.$$

A representation for the lower edge from ρ to is R is $z=re^{i2\pi}$ $(\rho\le r\le R)$. Hence the value of the integral of *f* along that edge from R to ρ is

$$-\int_\rho^R \frac{\exp\left[\frac{1}{3}(\ln r+i2\pi)\right]}{(r+a)(r+b)}dr=-e^{i2\pi/3}\int_\rho^R\frac{\sqrt[3]{r}}{(r+a)(r+b)}dr.$$

According to the residue theorem, then,

$$\int_\rho^R\frac{\sqrt[3]{r}}{(r+a)(r+b)}dr+\int_{C_R}f(z)dz-e^{i2\pi/3}\int_\rho^R\frac{\sqrt[3]{r}}{(r+a)(r+b)}dr+\int_{C_\rho}f(z)dz=2\pi i(B_1+B_2),$$

where

$$B_1 = \operatorname*{Res}_{z=-a} f(z) = \frac{\exp\left[\frac{1}{3}\log(-a)\right]}{-a+b} = -\frac{\exp\left[\frac{1}{3}(\ln a + i\pi)\right]}{a-b} = -\frac{e^{i\pi/3}\sqrt[3]{a}}{a-b}$$

and

$$B_2 = \operatorname*{Res}_{z=-b} f(z) = \frac{\exp\left[\frac{1}{3}\log(-b)\right]}{-b+a} = \frac{\exp\left[\frac{1}{3}(\ln b + i\pi)\right]}{-b+a} = \frac{e^{i\pi/3}\sqrt[3]{b}}{a-b}.$$

Consequently,

$$\left(1-e^{i2\pi/3}\right)\int_\rho^R \frac{\sqrt[3]{r}}{(r+a)(r+b)}dr = -\frac{2\pi i e^{i\pi/3}(\sqrt[3]{a}-\sqrt[3]{b})}{a-b} - \int_{C_\rho} f(z)dz - \int_{C_R} f(z)dz.$$

Now

$$\left|\int_{C_\rho} f(z)dz\right| \le \frac{\sqrt[3]{\rho}}{(a-\rho)(b-\rho)}2\pi\rho = \frac{2\pi\sqrt[3]{\rho}\,\rho}{(a-\rho)(b-\rho)} \to 0 \text{ as } \rho \to 0$$

and

$$\left|\int_{C_R} f(z)dz\right| \le \frac{\sqrt[3]{R}}{(R-a)(R-b)}2\pi R = \frac{2\pi R^2}{(R-a)(R-b)}\cdot\frac{1}{\sqrt[3]{R^2}} \to 0 \text{ as } R \to \infty.$$

Hence

$$\int_0^\infty \frac{\sqrt[3]{r}}{(r+a)(r+b)}dr = -\frac{2\pi i e^{i\pi/3}(\sqrt[3]{a}-\sqrt[3]{b})}{\left(1-e^{i2\pi/3}\right)(a-b)}\cdot\frac{e^{-i\pi/3}}{e^{-i\pi/3}} = \frac{2\pi i(\sqrt[3]{a}-\sqrt[3]{b})}{\left(e^{i\pi/3}-e^{-i\pi/3}\right)(a-b)}$$

$$= \frac{\pi(\sqrt[3]{a}-\sqrt[3]{b})}{\sin(\pi/3)(a-b)} = \frac{\pi(\sqrt[3]{a}-\sqrt[3]{b})}{\frac{\sqrt{3}}{2}(a-b)} = \frac{2\pi}{\sqrt{3}}\cdot\frac{\sqrt[3]{a}-\sqrt[3]{b}}{a-b}.$$

Replacing the variable of integration r here by x, we have the desired result:

$$\int_0^\infty \frac{\sqrt[3]{x}}{(x+a)(x+b)}dx = \frac{2\pi}{\sqrt{3}}\cdot\frac{\sqrt[3]{a}-\sqrt[3]{b}}{a-b} \qquad (a>b>0).$$

6. *(a)* Let us first use the branch

$$f(z)=\frac{z^{-1/2}}{z^2+1}=\frac{\exp\left(-\frac{1}{2}\log z\right)}{z^2+1}\qquad\left(|z|>0,-\frac{\pi}{2}<\arg z<\frac{3\pi}{2}\right)$$

and the indented path shown below to evaluate the improper integral

$$\int_0^\infty\frac{dx}{\sqrt{x}(x^2+1)}.$$

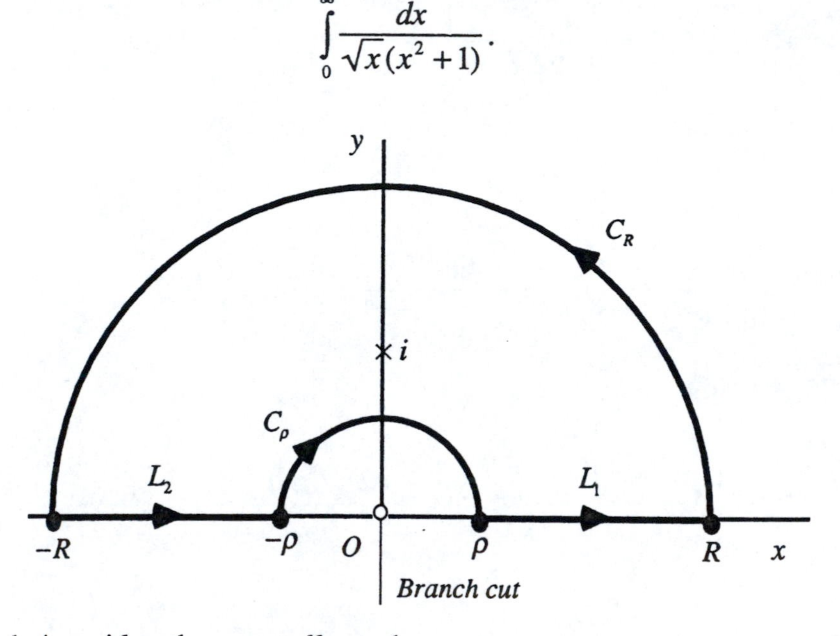

Cauchy's residue theorem tells us that

$$\int_{L_1}f(z)\,dz+\int_{C_R}f(z)\,dz+\int_{L_2}f(z)\,dz+\int_{C_\rho}f(z)\,dz=2\pi i\operatorname*{Res}_{z=i}f(z),$$

or

$$\int_{L_1}f(z)\,dz+\int_{L_2}f(z)\,dz=2\pi i\operatorname*{Res}_{z=i}f(z)-\int_{C_\rho}f(z)\,dz-\int_{C_R}f(z)\,dz.$$

Since

$$L_1\!: z=re^{i0}=r\ (\rho\le r\le R)\quad\text{and}\quad -L_2\!: z=re^{i\pi}=-r\ (\rho\le r\le R),$$

we may write

$$\int_{L_1}f(z)dz+\int_{L_2}f(z)dz=\int_\rho^R\frac{dr}{\sqrt{r}(r^2+1)}-i\int_\rho^R\frac{dr}{\sqrt{r}(r^2+1)}=(1-i)\int_\rho^R\frac{dr}{\sqrt{r}(r^2+1)}.$$

Thus

$$(1-i)\int_\rho^R\frac{dr}{\sqrt{r}(r^2+1)}=2\pi i\operatorname*{Res}_{z=i}f(z)-\int_{C_\rho}f(z)\,dz-\int_{C_R}f(z)\,dz.$$

Now the point $z = i$ is evidently a simple pole of $f(z)$, with residue

$$\operatorname*{Res}_{z=i} f(z) = \left.\frac{z^{-1/2}}{z+i}\right]_{z=i} = \frac{\exp\left[-\frac{1}{2}\log i\right]}{2i} = \frac{\exp\left[-\frac{1}{2}\left(\ln 1 + i\frac{\pi}{2}\right)\right]}{2i} = \frac{e^{-i\pi/4}}{2i} = \frac{1}{2i}\left(\frac{1-i}{\sqrt{2}}\right).$$

Furthermore,

$$\left|\int_{C_\rho} f(z)dz\right| \le \frac{\pi\rho}{\sqrt{\rho}(1-\rho^2)} = \frac{\pi\sqrt{\rho}}{1-\rho^2} \to 0 \text{ as } \rho \to 0$$

and

$$\left|\int_{C_R} f(z)dz\right| \le \frac{\pi\sqrt{R}}{(R^2-1)} = \frac{\pi}{\sqrt{R}\left(R - \frac{1}{R}\right)} \to 0 \text{ as } R \to \infty.$$

Finally, then, we have

$$(1-i)\int_0^\infty \frac{dr}{\sqrt{r}(r^2+1)} = \frac{\pi(1-i)}{\sqrt{2}},$$

which is the same as

$$\int_0^\infty \frac{dx}{\sqrt{x}(x^2+1)} = \frac{\pi}{\sqrt{2}}.$$

(b) To evaluate the improper integral $\int_0^\infty \frac{dx}{\sqrt{x}(x^2+1)}$, we now use the branch

$$f(z) = \frac{z^{-1/2}}{z^2+1} = \frac{\exp\left(-\frac{1}{2}\log z\right)}{z^2+1} \qquad (|z| > 0, 0 < \arg z < 2\pi)$$

and the simple closed contour shown in the figure below, which is similar to Fig. 99 in Sec. 77. We stipulate that $\rho < 1$ and $R > 1$, so that the singularities $z = \pm i$ are between C_ρ and C_R.

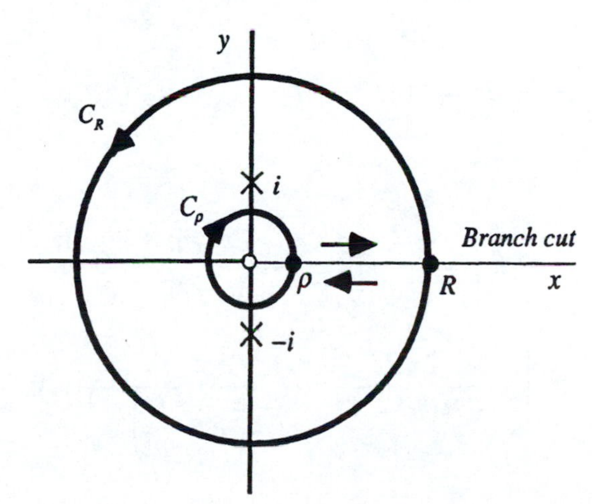

Since a parametric representation for the upper edge of the branch cut from ρ to R is $z = re^{i0}$ $(\rho \le r \le R)$, the value of the integral of f along that edge is

$$\int_\rho^R \frac{\exp\left[-\frac{1}{2}(\ln r + i0)\right]}{r^2+1}dr = \int_\rho^R \frac{1}{\sqrt{r}(r^2+1)}dr.$$

A representation for the lower edge from ρ to is R is $z = re^{i2\pi}$ $(\rho \le r \le R)$, and so the value of the integral of f along that edge from R to ρ is

$$-\int_\rho^R \frac{\exp\left[-\frac{1}{2}(\ln r + i2\pi)\right]}{r^2+1}dr = -e^{-i\pi}\int_\rho^R \frac{1}{\sqrt{r}(r^2+1)}dr = \int_\rho^R \frac{1}{\sqrt{r}(r^2+1)}dr.$$

Hence, by the residue theorem,

$$\int_\rho^R \frac{1}{\sqrt{r}(r^2+1)}dr + \int_{C_R} f(z)dz + \int_\rho^R \frac{1}{\sqrt{r}(r^2+1)}dr + \int_{C_\rho} f(z)dz = 2\pi i(B_1 + B_2),$$

where

$$B_1 = \operatorname*{Res}_{z=i} f(z) = \left.\frac{z^{-1/2}}{z+i}\right]_{z=i} = \frac{\exp\left[-\frac{1}{2}\log i\right]}{2i} = \frac{\exp\left[-\frac{1}{2}\left(\ln 1 + i\frac{\pi}{2}\right)\right]}{2i} = \frac{e^{-i\pi/4}}{2i}$$

and

$$B_2 = \operatorname*{Res}_{z=-i} f(z) = \left.\frac{z^{-1/2}}{z-i}\right]_{z=-i} = \frac{\exp\left[-\frac{1}{2}\log(-i)\right]}{-2i} = \frac{\exp\left[-\frac{1}{2}\left(\ln 1 + i\frac{3\pi}{2}\right)\right]}{-2i} = -\frac{e^{-i3\pi/4}}{2i}.$$

That is,

$$2\int_\rho^R \frac{1}{\sqrt{r}(r^2+1)}dr = \pi(e^{-i\pi/4} - e^{-i3\pi/4}) - \int_{C_\rho} f(z)dz - \int_{C_R} f(z)dz.$$

Since

$$\left|\int_{C_\rho} f(z)dz\right| \le \frac{2\pi\rho}{\sqrt{\rho}(1-\rho^2)} = \frac{2\pi\sqrt{\rho}}{1-\rho^2} \to 0 \text{ as } \rho \to 0$$

and

$$\left|\int_{C_R} f(z)dz\right| \le \frac{2\pi R}{\sqrt{R}(R^2-1)} = \frac{2\pi}{\sqrt{R}\left(R - \frac{1}{R}\right)} \to 0 \text{ as } R \to \infty,$$

we now find that

$$\int_0^\infty \frac{1}{\sqrt{r}(r^2+1)}dr = \pi\frac{e^{-i\pi/4}-e^{-i3\pi/4}}{2} = \pi\frac{e^{-i\pi/4}+e^{-i3\pi/4}e^{i\pi}}{2}$$

$$= \pi\frac{e^{i\pi/4}+e^{-i\pi/4}}{2} = \pi\cos\left(\frac{\pi}{4}\right) = \frac{\pi}{\sqrt{2}}.$$

When x, instead of r, is used as the variable of integration here, we have the desired result:

$$\int_0^\infty \frac{dx}{\sqrt{x}(x^2+1)} = \frac{\pi}{\sqrt{2}}.$$

SECTION 78

1. Write

$$\int_0^{2\pi} \frac{d\theta}{5+4\sin\theta} = \int_C \frac{1}{5+4\left(\dfrac{z-z^{-1}}{2i}\right)}\cdot\frac{dz}{iz} = \int_C \frac{dz}{2z^2+5iz-2},$$

where C is the positively oriented unit circle $|z|=1$. The quadratic formula tells us that the singular points of the integrand on the far right here are $z=-i/2$ and $z=-2i$. The point $z=-i/2$ is a simple pole interior to C; and the point $z=-2i$ is exterior to C. Thus

$$\int_0^{2\pi} \frac{d\theta}{5+4\sin\theta} = 2\pi i \operatorname*{Res}_{z=-i/2}\left[\frac{1}{2z^2+5iz-2}\right] = 2\pi i\left[\frac{1}{4z+5i}\right]_{z=-i/2} = 2\pi i\left(\frac{1}{3i}\right) = \frac{2\pi}{3}.$$

2. To evaluate the definite integral in question, write

$$\int_{-\pi}^{\pi} \frac{d\theta}{1+\sin^2\theta} = \int_C \frac{1}{1+\left(\dfrac{z-z^{-1}}{2i}\right)^2}\cdot\frac{dz}{iz} = \int_C \frac{4iz\,dz}{z^4-6z^2+1},$$

where C is the positively oriented unit circle $|z|=1$. This circle is shown below.

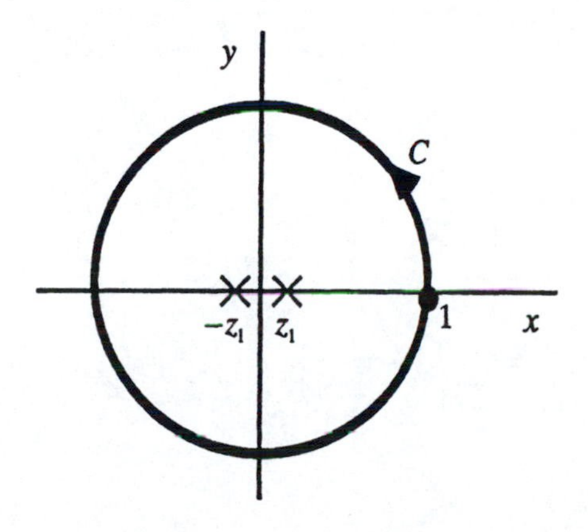

Solving the equation $(z^2)^2-6(z^2)+1=0$ for z^2 with the aid of the quadratic formula, we find that the zeros of the polynomial z^4-6z^2+1 are the numbers z such that $z^2=3\pm 2\sqrt{2}$. Those zeros are, then, $z=\pm\sqrt{3+2\sqrt{2}}$ and $z=\pm\sqrt{3-2\sqrt{2}}$. The first two of these zeros are exterior to the circle, and the second two are inside of it. So the singularities of the integrand in our contour integral are

$$z_1=\sqrt{3-2\sqrt{2}} \quad \text{and} \quad z_2=-z_1,$$

indicated in the figure. This means that

$$\int_{-\pi}^{\pi}\frac{d\theta}{1+\sin^2\theta}=2\pi i(B_1+B_2),$$

where

$$B_1=\operatorname*{Res}_{z=z_1}\frac{4iz}{z^4-6z^2+1}=\frac{4iz_1}{4z_1^3-12z_1}=\frac{i}{z_1^2-3}=\frac{i}{(3-2\sqrt{2})-3}=-\frac{i}{2\sqrt{2}}$$

and

$$B_2=\operatorname*{Res}_{z=-z_1}\frac{4iz}{z^4-6z^2+1}=\frac{-4iz_1}{-4z_1^3+12z_1}=\frac{i}{z_1^2-3}=-\frac{i}{2\sqrt{2}}.$$

Since

$$2\pi i(B_1+B_2)=2\pi i\left(-\frac{i}{\sqrt{2}}\right)=\frac{2\pi}{\sqrt{2}}\cdot\frac{\sqrt{2}}{\sqrt{2}}=\sqrt{2}\pi,$$

the desired result is

$$\int_{-\pi}^{\pi}\frac{d\theta}{1+\sin^2\theta}=\sqrt{2}\pi.$$

7. Let C be the positively oriented unit circle $|z|=1$. In view of the binomial formula (Sec. 3)

$$\int_0^{\pi}\sin^{2n}\theta\,d\theta=\frac{1}{2}\int_{-\pi}^{\pi}\sin^{2n}\theta\,d\theta=\frac{1}{2}\int_C\left(\frac{z-z^{-1}}{2i}\right)^{2n}\frac{dz}{iz}=\frac{1}{2^{2n+1}(-1)^n i}\int_C\frac{(z-z^{-1})^{2n}}{z}dz$$

$$=\frac{1}{2^{2n+1}(-1)^n i}\int_C\sum_{k=0}^{n}\binom{2n}{k}z^{2n-k}(-z^{-1})^k z^{-1}dz$$

$$=\frac{1}{2^{2n+1}(-1)^n i}\sum_{k=0}^{n}\binom{2n}{k}(-1)^k\int_C z^{2n-2k-1}\,dz.$$

Now each of these last integrals has value zero except when $k=n$:

$$\int_C z^{-1}dz = 2\pi i.$$

Consequently,

$$\int_0^\pi \sin^{2n}\theta\, d\theta = \frac{1}{2^{2n+1}(-1)^n i}\cdot\frac{(2n)!(-1)^n 2\pi i}{(n!)^2} = \frac{(2n)!}{2^{2n}(n!)^2}\pi.$$

SECTION 80

5. We are given a function f that is analytic inside and on a positively oriented simple closed contour C, and we assume that f has no zeros on C. Also, f has n zeros z_k $(k=1,2,\ldots,n)$ inside C, where each z_k is of multiplicity m_k. (See the figure below.)

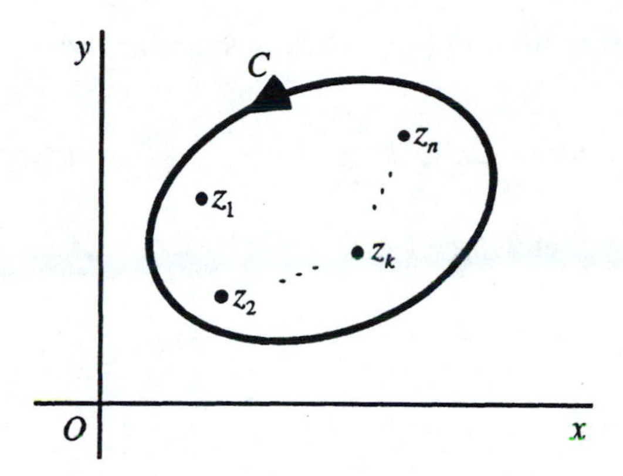

The object here is to show that

$$\int_C \frac{zf'(z)}{f(z)}dz = 2\pi i\sum_{k=1}^{n} m_k z_k.$$

To do this, we consider the kth zero and start with the fact that

$$f(z) = (z-z_k)^{m_k} g(z),$$

where $g(z)$ is analytic and nonzero at z_k. From this, it is straightforward to show that

$$\frac{zf'(z)}{f(z)} = \frac{m_k z}{z-z_k} + \frac{zg'(z)}{g(z)} = \frac{m_k(z-z_k)+m_k z_k}{z-z_k} + \frac{zg'(z)}{g(z)} = m_k + \frac{zg'(z)}{g(z)} + \frac{m_k z_k}{z-z_k}.$$

Since the term $\dfrac{zg'(z)}{g(z)}$ here has a Taylor series representation at z_k, it follows that $\dfrac{zf'(z)}{f(z)}$ has a simple pole at z_k and that

$$\operatorname*{Res}_{z=z_k}\frac{zf'(z)}{f(z)} = m_k z_k.$$

An application of the residue theorem now yields the desired result.

6. *(a)* To determine the number of zeros of the polynomial $z^6 - 5z^4 + z^3 - 2z$ inside the circle $|z| = 1$, we write

$$f(z) = -5z^4 \quad \text{and} \quad g(z) = z^6 + z^3 - 2z.$$

We then observe that when z is on the circle,

$$|f(z)| = 5 \quad \text{and} \quad |g(z)| \le |z|^6 + |z|^3 + 2|z| = 4.$$

Since $|f(z)| > |g(z)|$ on the circle and since $f(z)$ has 4 zeros, counting multiplicities, inside it, the theorem in Sec. 80 tells is that the sum

$$f(z) + g(z) = z^6 - 5z^4 + z^3 - 2z$$

also has four zeros, counting multiplicities, inside the circle.

(b) Let us write the polynomial $2z^4 - 2z^3 + 2z^2 - 2z + 9$ as the sum $f(z) + g(z)$, where

$$f(z) = 9 \quad \text{and} \quad g(z) = 2z^4 - 2z^3 + 2z^2 - 2z.$$

Observe that when z is on the circle $|z| = 1$,

$$|f(z)| = 9 \quad \text{and} \quad |g(z)| \le 2|z|^4 + 2|z|^3 + 2|z|^2 + 2|z| = 8.$$

Since $|f(z)| > |g(z)|$ on the circle and since $f(z)$ has no zeros inside it, the sum $f(z) + g(z) = 2z^4 - 2z^3 + 2z^2 - 2z + 9$ has no zeros there either.

7. Let C denote the circle $|z| = 2$.

(a) The polynomial $z^4 + 3z^3 + 6$ can be written as the sum of the polynomials

$$f(z) = 3z^3 \quad \text{and} \quad g(z) = z^4 + 6.$$

On C,

$$|f(z)| = 3|z|^3 = 24 \quad \text{and} \quad |g(z)| = |z^4 + 6| \le |z|^4 + 6 = 22.$$

Since $|f(z)| > |g(z)|$ on C and $f(z)$ has 3 zeros, counting multiplicities, inside C, it follows that the original polynomial has 3 zeros, counting multiplicities, inside C.

(b) The polynomial $z^4 - 2z^3 + 9z^2 + z - 1$ can be written as the sum of the polynomials

$$f(z) = 9z^2 \quad \text{and} \quad g(z) = z^4 - 2z^3 + z - 1.$$

On C,

$$|f(z)| = 9|z|^2 = 36 \quad \text{and} \quad |g(z)| = |z^4 - 2z^3 + z - 1| \le |z|^4 + 2|z|^3 + |z| + 1 = 35.$$

Since $|f(z)|>|g(z)|$ on C and $f(z)$ has 2 zeros, counting multiplicities, inside C, it follows that the original polynomial has 2 zeros, counting multiplicities, inside C.

(c) The polynomial $z^5+3z^3+z^2+1$ can be written as the sum of the polynomials

$$f(z)=z^5 \quad \text{and} \quad g(z)=3z^3+z^2+1.$$

On C,

$$|f(z)|=|z|^5=32 \quad \text{and} \quad |g(z)|=|3z^3+z^2+1|\le 3|z|^3+|z|^2+1=29.$$

Since $|f(z)|>|g(z)|$ on C and $f(z)$ has 5 zeros, counting multiplicities, inside C, it follows that the original polynomial has 5 zeros, counting multiplicities, inside C.

10. The problem here is to give an alternative proof of the fact that any polynomial

$$P(z)=a_0+a_1z+\cdots+a_{n-1}z^{n-1}+a_nz^n \qquad (a_n\neq 0),$$

where $n\ge 1$, has precisely n zeros, counting multiplicities. Without loss of generality, we may take $a_n=1$ since

$$P(z)=a_n\left(\frac{a_0}{a_n}+\frac{a_1}{a_n}z+\cdots+\frac{a_{n-1}}{a_n}z^{n-1}+z^n\right).$$

Let

$$f(z)=z^n \quad \text{and} \quad g(z)=a_0+a_1z+\cdots+a_{n-1}z^{n-1}.$$

Then let R be so large that

$$R>1+|a_0|+|a_1|+\cdots+|a_{n-1}|.$$

If z is a point on the circle $C: |z|=R$, we find that

$$\begin{aligned}|g(z)|&\le |a_0|+|a_1||z|+\cdots+|a_{n-1}||z|^{n-1}=|a_0|+|a_1|R+\cdots+|a_{n-1}|R^{n-1}\\ &<|a_0|R^{n-1}+|a_1|R^{n-1}+\cdots+|a_{n-1}|R^{n-1}=(|a_0|+|a_1|+\cdots+|a_{n-1}|)R^{n-1}\\ &<RR^{n-1}=R^n=|z|^n=|f(z)|.\end{aligned}$$

Since $f(z)$ has precisely n zeros, counting multiplicities, inside C and since R can be made arbitrarily large, the desired result follows.

1. The singularities of the function

$$F(s)=\frac{2s^3}{s^4-4}$$

are the fourth roots of 4. They are readily found to be

$$s=\sqrt{2}\,e^{ik\pi/2} \qquad (k=0,1,2,3),$$

or

$$\sqrt{2},\quad \sqrt{2}i,\quad -\sqrt{2},\quad \text{and}\quad -\sqrt{2}i.$$

See the figure below, where $\gamma>\sqrt{2}$ and $R>\sqrt{2}+\gamma$.

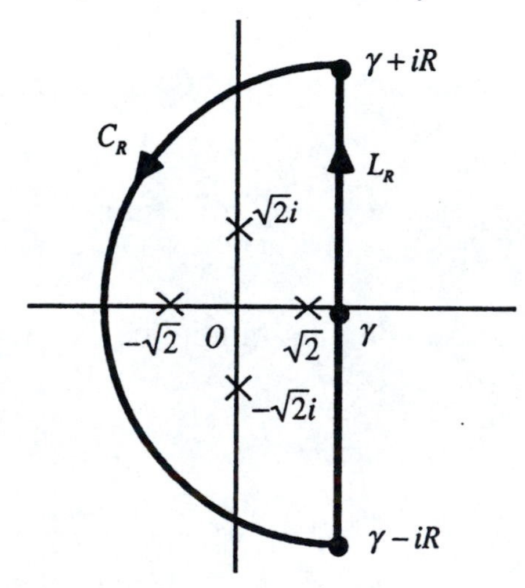

The function

$$e^{st}F(s)=\frac{2s^3e^{st}}{s^4-4}$$

has simple poles at the points

$$s_0=\sqrt{2},\quad s_1=\sqrt{2}i,\quad s_2=-\sqrt{2},\quad \text{and}\quad s_3=-\sqrt{2}i;$$

and

$$\sum_{n=0}^{3}\operatorname*{Res}_{s=s_n}\left[e^{st}F(s)\right]=\sum_{n=0}^{3}\operatorname*{Res}_{s=s_n}\frac{2s^3e^{st}}{s^4-4}=\sum_{n=0}^{3}\frac{2s_n^3e^{s_nt}}{4s_n^3}=\sum_{n=0}^{3}\frac{1}{2}e^{s_nt}$$

$$=\frac{1}{2}e^{\sqrt{2}t}+\frac{1}{2}e^{i\sqrt{2}t}+\frac{1}{2}e^{-\sqrt{2}t}+\frac{1}{2}e^{-i\sqrt{2}t}$$

$$=\frac{e^{\sqrt{2}t}+e^{-\sqrt{2}t}}{2}+\frac{e^{i\sqrt{2}t}+e^{-i\sqrt{2}t}}{2}$$

$$=\cosh\sqrt{2}t+\cos\sqrt{2}t.$$

Suppose now that s is a point on C_R, and observe that

$$|s|=|\gamma+Re^{i\theta}|\le\gamma+R=R+\gamma \quad\text{and}\quad |s|=|\gamma+Re^{i\theta}|\ge|\gamma-R|=R-\gamma>\sqrt{2}.$$

It follows that

$$|2s^3|=2|s|^3\le 2(R+\gamma)^3$$

and

$$|s^4-4|\ge||s|^4-4|\ge(R-\gamma)^4-4>0.$$

Consequently,

$$|F(s)|\le\frac{2(R+\gamma)^3}{(R-\gamma)^4-4}\to 0 \text{ as } R\to\infty.$$

This ensures that

$$f(t)=\cosh\sqrt{2}t+\cos\sqrt{2}t.$$

2. The polynomials in the denominator of

$$F(s)=\frac{2s-2}{(s+1)(s^2+2s+5)}$$

have zeros at $s=-1$ and $s=-1\pm 2i$. Let us, then, write

$$e^{st}F(s)=\frac{e^{st}(2s-2)}{(s+1)(s-s_1)(s-\overline{s}_1)},$$

where $s_1=-1+2i$. The points -1, s_1, and $\overline{s}_1$ are evidently simple poles of $e^{st}F(s)$ with the following residues:

$$B_1=\operatorname*{Res}_{z=-1}\left[e^{st}F(s)\right]=\left.\frac{e^{st}(2s-2)}{(s-s_1)(s-\overline{s}_1)}\right]_{s=-1}=-e^{-t},$$

$$B_2=\operatorname*{Res}_{s=s_1}\left[e^{st}F(s)\right]=\frac{e^{s_1t}(2s_1-2)}{(s_1+1)(s_1-\overline{s}_1)}=\left(\frac{1}{2}-\frac{i}{2}\right)e^{-t}e^{i2t},$$

$$B_3=\operatorname*{Res}_{s=\overline{s}_1}\left[e^{st}F(s)\right]=\frac{e^{\overline{s}_1t}(2\overline{s}_1-2)}{(\overline{s}_1+1)(\overline{s}_1-s_1)}=\overline{\left[\frac{e^{s_1t}(2s_1-2)}{(s_1+1)(s_1-\overline{s}_1)}\right]}=\overline{B}_2=\left(\frac{1}{2}+\frac{i}{2}\right)e^{-t}e^{-i2t}.$$

It is easy to see that

$$B_1 + B_2 + B_3 = -e^{-t} + \left(\frac{1}{2} - \frac{i}{2}\right)e^{-t}e^{i2t} + \left(\frac{1}{2} + \frac{i}{2}\right)e^{-t}e^{-i2t}$$

$$= -e^{-t} + e^{-t}\left(\frac{e^{i2t} - e^{-i2t}}{2i} + \frac{e^{i2t} + e^{-i2t}}{2}\right) = e^{-t}(\sin 2t + \cos 2t - 1).$$

Now let s be any point on the semicircle shown below, where $\gamma > 0$ and $R > \sqrt{5} + \gamma$.

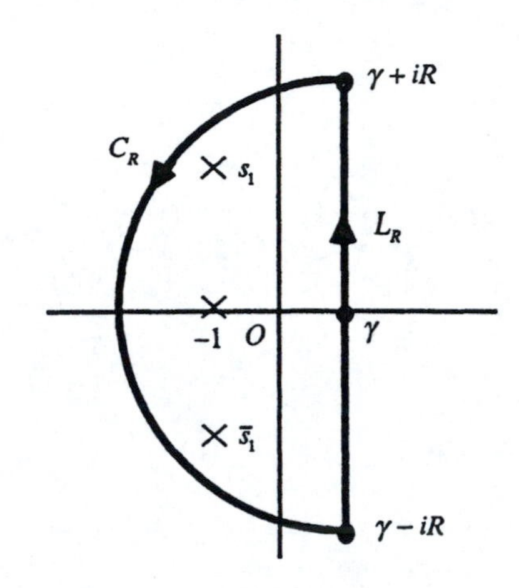

Since

$$|s| = |\gamma + Re^{i\theta}| \leq \gamma + R = R + \gamma \quad \text{and} \quad |s| = |\gamma + Re^{i\theta}| \geq |\gamma - R| = R - \gamma > \sqrt{5},$$

we find that

$$|2s - 2| \leq 2|s| + 2 \leq 2(R + \gamma) + 2,$$

$$|s + 1| \geq ||s| - 1| \geq (R - \gamma) - 1 > 0,$$

and

$$|s^2 + 2s + 5| = |s - s_1||s - \bar{s}_1| \geq (|s| - |s_1|)^2 \geq \left[(R - \gamma)^2 - \sqrt{5}\right]^2 > 0.$$

Thus

$$|F(s)| = \frac{|2s - 2|}{|s + 1||s^2 + 2s + 5|} \leq \frac{2(R + \gamma) + 2}{[(R - \gamma) - 1]\left[(R - \gamma)^2 - \sqrt{5}\right]^2} \to 0 \text{ as } R \to \infty,$$

and we may conclude that

$$f(t) = e^{-t}(\sin 2t + \cos 2t - 1).$$

4. The function

$$F(s) = \frac{s^2 - a^2}{(s^2 + a^2)^2} \qquad (a > 0)$$

has singularities at $s = \pm ai$. So we consider the simple closed contour shown below, where $\gamma > 0$ and $R > a + \gamma$.

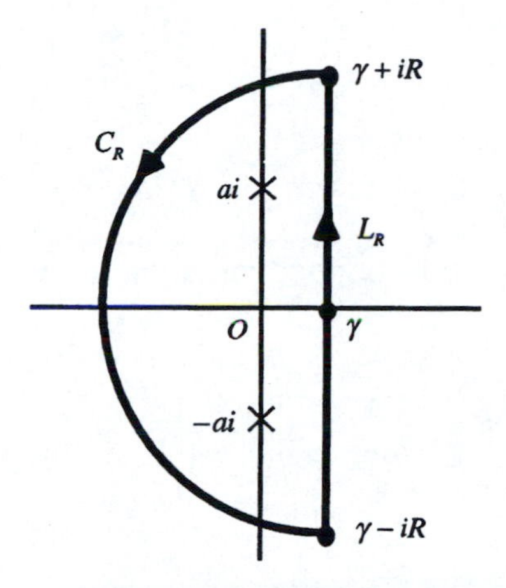

Upon writing

$$F(s) = \frac{\phi(s)}{(s - ai)^2} \quad \text{where} \quad \phi(s) = \frac{s^2 - a^2}{(s + ai)^2},$$

we see that $\phi(s)$ is analytic and nonzero at $s_0 = ai$. Hence s_0 is a pole of order $m = 2$ of $F(s)$. Furthermore, $\overline{F(s)} = F(\overline{s})$ at points where $F(s)$ is analytic. Consequently, $\overline{s}_0$ is also a pole of order 2 of $F(s)$; and we know from expression (2), Sec. 82, that

$$\operatorname*{Res}_{s=s_0}\left[e^{st}F(s)\right] + \operatorname*{Res}_{s=\overline{s}_0}\left[e^{st}F(s)\right] = 2\operatorname{Re}\left[e^{iat}(b_1 + b_2 t)\right],$$

where b_1 and b_2 are the coefficients in the principal part

$$\frac{b_1}{s - ai} + \frac{b_2}{(s - ai)^2}$$

of $F(s)$ at ai. These coefficients are readily found with the aid of the first two terms in the Taylor series for $\phi(s)$ about $s_0 = ai$:

$$F(s) = \frac{1}{(s - ai)^2}\phi(s) = \frac{1}{(s - ai)^2}\left[\phi(ai) + \frac{\phi'(ai)}{1!}(s - ai) + \cdots\right]$$

$$= \frac{\phi(ai)}{(s-ai)^2} + \frac{\phi'(ai)}{s-ai} + \cdots \qquad (0 < |s-ai| < 2a).$$

It is straightforward to show that $\phi(ai) = 1/2$ and $\phi'(ai) = 0$, and we find that $b_1 = 0$ and $b_2 = 1/2$. Hence

$$\operatorname*{Res}_{s=s_0}\left[e^{st}F(s)\right] + \operatorname*{Res}_{s=\bar{s}_0}\left[e^{st}F(s)\right] = 2\operatorname{Re}\left[e^{iat}\left(\frac{1}{2}t\right)\right] = t\cos at.$$

We can, then, conclude that

$$f(t) = t\cos at \qquad (a > 0),$$

provided that $F(s)$ satisfies the desired boundedness condition. As for that condition, when z is a point on C_R,

$$|z| = |\gamma + Re^{i\theta}| \le \gamma + R = R + \gamma \quad \text{and} \quad |z| = |\gamma + Re^{i\theta}| \ge |\gamma - R| = R - \gamma > a;$$

and this means that

$$|z^2 - a^2| \le |z|^2 + a^2 \le (R+\gamma)^2 + a^2 \quad \text{and} \quad |z^2 + a^2| \ge ||z|^2 - a^2| \ge (R-\gamma)^2 - a^2 > 0.$$

Hence

$$|F(z)| \le \frac{(R+\gamma)^2 + a^2}{[(R-\gamma)^2 - a^2]^2} \to 0 \text{ as } R \to \infty.$$

6. We are given

$$F(s) = \frac{\sinh(xs)}{s^2 \cosh s} \qquad (0 < x < 1),$$

which has isolated singularities at the points

$$s_0 = 0, \quad s_n = \frac{(2n-1)\pi}{2}i, \quad \text{and} \quad \bar{s}_n = -\frac{(2n-1)\pi}{2}i \qquad (n = 1, 2, \ldots).$$

This function has the property $\overline{F(s)} = F(\bar{s})$, and so

$$f(t) = \operatorname*{Res}_{s=s_0}\left[e^{st}F(s)\right] + \sum_{n=1}^{\infty}\left\{\operatorname*{Res}_{s=s_n}\left[e^{st}F(s)\right] + \operatorname*{Res}_{s=\bar{s}_n}\left[e^{st}F(s)\right]\right\}.$$

To find the residue at $s_0 = 0$, we write

$$\frac{\sinh(xs)}{s^2\cosh s} = \frac{xs + (xs)^3/3! + \cdots}{s^2\left(1 + s^2/2! + \cdots\right)} = \frac{x + x^3s^2/6 + \cdots}{s + s^3/2 + \cdots} \qquad \left(0 < |s| < \frac{\pi}{2}\right).$$

Division of series then reveals that s_0 is a simple pole of $F(s)$, with residue x; and, according to expression (3), Sec. 82,

$$\operatorname*{Res}_{s=s_0}\left[e^{st}F(s)\right]=\operatorname*{Res}_{s=s_0}F(s)=x.$$

As for the residues of $F(s)$ at the singular points s_n $(n=1,2,\ldots)$, we write

$$F(s)=\frac{p(s)}{q(s)} \quad \text{where} \quad p(s)=\sinh(xs) \text{ and } q(s)=s^2\cosh s.$$

We note that

$$p(s_n)=i\sin\frac{(2n-1)\pi x}{2}\neq 0 \quad \text{and} \quad q(s_n)=0;$$

furthermore, since

$$q'(s)=2s\cosh s+s^2\sinh s,$$

we find that

$$q'(s_n)=-\frac{(2n-1)^2\pi^2}{4}i\sin\frac{(2n-1)\pi}{2}=-i\frac{(2n-1)^2\pi^2}{4}\sin\left(n\pi-\frac{\pi}{2}\right)$$

$$=-i\frac{(2n-1)^2\pi^2}{4}\left(\sin n\pi\cos\frac{\pi}{2}-\cos n\pi\sin\frac{\pi}{2}\right)=\frac{(2n-1)^2\pi^2}{4}(-1)^n i\neq 0.$$

In view of Theorem 2 in Sec. 69, then, s_n is a simple pole of $F(s)$, and

$$\operatorname*{Res}_{s=s_n}F(s)=\frac{p(s_n)}{q'(s_n)}=\frac{4}{\pi^2}\cdot\frac{(-1)^n}{(2n-1)^2}\sin\frac{(2n-1)\pi x}{2}.$$

Expression (4), Sec. 82, now gives us

$$\operatorname*{Res}_{s=s_n}\left[e^{st}F(s)\right]+\operatorname*{Res}_{s=\overline{s}_n}\left[e^{st}F(s)\right]=2\operatorname{Re}\left\{\frac{4}{\pi^2}\cdot\frac{(-1)^n}{(2n-1)^2}\sin\frac{(2n-1)\pi x}{2}\exp\left[i\frac{(2n-1)\pi t}{2}\right]\right\}$$

$$=\frac{8}{\pi^2}\cdot\frac{(-1)^n}{(2n-1)^2}\sin\frac{(2n-1)\pi x}{2}\cos\frac{(2n-1)\pi t}{2}.$$

Summing all of the above residues, we arrive at the final result:

$$f(t) = x + \frac{8}{\pi^2}\sum_{n=1}^{\infty}\frac{(-1)^n}{(2n-1)^2}\sin\frac{(2n-1)\pi x}{2}\cos\frac{(2n-1)\pi t}{2}.$$

7. The function

$$F(s) = \frac{1}{s\cosh(s^{1/2})},$$

where it is agreed that the branch cut of $s^{1/2}$ does not lie along the negative real axis, has isolated singularities at $s_0 = 0$ and when $\cosh(s^{1/2}) = 0$, or at the points $s_n = -\frac{(2n-1)^2\pi^2}{4}$ $(n = 1, 2, \ldots)$. The point s_0 is a simple pole of $F(s)$, as is seen by writing

$$\frac{1}{s\cosh(s^{1/2})} = \frac{1}{s\left[1 + (s^{1/2})^2/2! + (s^{1/2})^4/4! + \cdots\right]} = \frac{1}{s + s^2/2 + s^3/24 + \cdots}$$

and dividing this last denominator into 1. In fact, the residue is found to be 1; and expression (3), Sec. 82, tells us that

$$\operatorname*{Res}_{s=s_0}\left[e^{st}F(s)\right] = \operatorname*{Res}_{s=s_0} F(s) = 1.$$

As for the other singularities, we write

$$F(s) = \frac{p(s)}{q(s)} \quad \text{where} \quad p(s) = 1 \text{ and } q(s) = s\cosh(s^{1/2}).$$

Now

$$p(s_n) = 1 \neq 0 \quad \text{and} \quad q(s_n) = 0;$$

also, since

$$q'(s) = \frac{1}{2}s^{1/2}\sinh(s^{1/2}) + \cosh(s^{1/2}),$$

it is straightforward to show that

$$q'(s_n) = -\frac{(2n-1)\pi}{4}\sin\left(n\pi - \frac{\pi}{2}\right) = \frac{(2n-1)\pi}{4}(-1)^n \neq 0.$$

So each point s_n is a simple pole of $F(s)$, and

$$\operatorname*{Res}_{s=s_n} F(s) = \frac{p(s_n)}{q'(s_n)} = \frac{4}{\pi}\cdot\frac{(-1)^n}{2n-1}.$$

Consequently, according to expression (3), Sec. 82,

$$\operatorname*{Res}_{s=s_n}\left[e^{st}F(s)\right] = e^{s_n t}\operatorname*{Res}_{s=s_n} F(s) = \frac{4}{\pi}\cdot\frac{(-1)^n}{2n-1}\exp\left[-\frac{(2n-1)^2\pi^2 t}{4}\right] \qquad (n=1,2,\ldots).$$

Finally, then,

$$f(t) = \operatorname*{Res}_{s=s_0}\left[e^{st}F(s)\right] + \sum_{n=1}^{\infty}\operatorname*{Res}_{s=s_n}\left[e^{st}F(s)\right],$$

or

$$f(t) = 1 + \frac{4}{\pi}\sum_{n=1}^{\infty}\frac{(-1)^n}{2n-1}\exp\left[-\frac{(2n-1)^2\pi^2 t}{4}\right].$$

8. Here we are given the function

$$F(s) = \frac{\coth(\pi s/2)}{s^2+1} = \frac{\cosh(\pi s/2)}{(s^2+1)\sinh(\pi s/2)},$$

which has the property $\overline{F(s)} = F(\overline{s})$. We consider first the singularities at $s=\pm i$. Upon writing

$$F(s) = \frac{\phi(s)}{s-i} \quad \text{where} \quad \phi(s) = \frac{\cosh(\pi s/2)}{(s+i)\sinh(\pi s/2)},$$

we find that, since $\phi(i)=0$, the point i is a removable singularity of $F(s)$ [see Exercise 3(*b*), Sec. 65]; and the same is true of the point $-i$. At each of these points, it follows that the residue of $e^{st}F(s)$ is 0. The other singularities occur when $\pi s/2 = n\pi i$ $(n=0,\pm1,\pm2,\ldots)$, or at the points $s=2ni$ $(n=0,\pm1,\pm2,\ldots)$. To find the residues, we write

$$F(s) = \frac{p(s)}{q(s)} \quad \text{where} \quad p(s) = \cosh\left(\frac{\pi s}{2}\right) \quad \text{and} \quad q(s) = (s^2+1)\sinh\left(\frac{\pi s}{2}\right)$$

and note that

$$p(2ni) = \cosh(n\pi i) = \cos(n\pi) = (-1)^n \neq 0 \quad \text{and} \quad q(2ni) = 0.$$

Furthermore, since

$$q'(s) = (s^2+1)\frac{\pi}{2}\cosh\left(\frac{\pi s}{2}\right) + 2s\sinh\left(\frac{\pi s}{2}\right),$$

we have

$$q'(2ni) = (-4n^2+1)\frac{\pi}{2}\cosh(n\pi i) = (-4n^2+1)\frac{\pi}{2}\cos(n\pi) = -\frac{\pi(4n^2-1)}{2}(-1)^n \neq 0.$$

Thus

$$\operatorname*{Res}_{s=2ni} F(s) = \frac{p(2ni)}{q'(2ni)} = -\frac{2}{\pi}\cdot\frac{1}{4n^2-1} \qquad (n = 0, \pm 1, \pm 2, \ldots).$$

Expressions (3) and (4) in Sec. 82 now tell us that

$$\operatorname*{Res}_{s=0}\left[e^{st}F(s)\right] = \operatorname*{Res}_{s=0} F(s) = \frac{2}{\pi}$$

and

$$\operatorname*{Res}_{s=2ni}\left[e^{st}F(s)\right] + \operatorname*{Res}_{s=-2ni}\left[e^{st}F(s)\right] = 2\operatorname{Re}\left[e^{i2nt}\left(-\frac{2}{\pi}\cdot\frac{1}{4n^2-1}\right)\right] = -\frac{4}{\pi}\cdot\frac{\cos 2nt}{4n^2-1} \qquad (n = 1, 2, \ldots).$$

The desired function of t is, then,

$$f(t) = \frac{2}{\pi} - \frac{4}{\pi}\sum_{n=1}^{\infty}\frac{\cos 2nt}{4n^2-1}.$$

9. The function

$$F(s) = \frac{\sinh(xs^{1/2})}{s^2\sinh(s^{1/2})} \qquad (0 < x < 1),$$

where it is agreed that the branch cut of $s^{1/2}$ does not lie along the negative real axis, has isolated singularities at $s = 0$ and when $\sinh(s^{1/2}) = 0$, or at the points $s = -n^2\pi^2$ $(n = 1, 2, \ldots)$. The point $s = 0$ is a pole of order 2 of $F(s)$, as is seen by first writing

$$\frac{\sinh(xs^{1/2})}{s^2\sinh(s^{1/2})} = \frac{xs^{1/2} + (xs^{1/2})^3/3! + (xs^{1/2})^5/5! + \cdots}{s^2\left[s^{1/2} + (s^{1/2})^3/3! + (s^{1/2})^5/5! + \cdots\right]} = \frac{x + x^3 s/6 + x^5 s^2/120 + \cdots}{s^2 + s^3/6 + s^4/120 + \cdots}$$

and dividing the series in the denominator into the series in the numerator. The result is

$$\frac{\sinh(xs^{1/2})}{s^2\sinh(s^{1/2})} = x\frac{1}{s^2} + \frac{1}{6}(x^3 - x)\frac{1}{s} + \cdots \qquad (0 < |s| < \pi^2).$$

In view of expression (1), Sec. 82, then,

$$\operatorname*{Res}_{s=0}\left[e^{st}F(s)\right] = \frac{1}{6}(x^3 - x) + xt = \frac{1}{6}x(x^2 - 1) + xt.$$

As for the singularities $s = -n^2\pi^2$ $(n = 1, 2, \ldots)$, we write

$$F(s) = \frac{p(s)}{q(s)} \quad \text{where} \quad p(s) = \sinh(xs^{1/2}) \text{ and } q(s) = s^2\sinh(s^{1/2}).$$

Observe that $p(-n^2\pi^2) \neq 0$ and $q(-n^2\pi^2) = 0$. Also, since

$$q'(s) = 2s\sinh(s^{1/2}) + \frac{1}{2}s\,s^{1/2}\cosh(s^{1/2}),$$

it is easy to see that $q'(-n^2\pi^2) \neq 0$. So the points $s = -n^2\pi^2$ $(n = 1, 2, \ldots)$, are simple poles of $F(s)$, and

$$\operatorname*{Res}_{s=-n^2\pi^2} F(s) = \left.\frac{p(s)}{q'(s)}\right]_{s=-n^2\pi^2} = \left.\frac{2\sinh(xs^{1/2})}{s\,s^{1/2}\cosh(s^{1/2})}\right]_{s=-n^2\pi^2} = \frac{2}{\pi^3}\cdot\frac{(-1)^{n+1}}{n^3}\sin n\pi x \qquad (n = 1, 2, \ldots).$$

Thus, in view of expression (3), Sec. 82,

$$\operatorname*{Res}_{s=-n^2\pi^2}\left[e^{st}F(s)\right] = \frac{2}{\pi^3}\cdot\frac{(-1)^{n+1}}{n^3}e^{-n^2\pi^2 t}\sin n\pi x \qquad (n = 1, 2, \ldots).$$

Finally, since

$$f(t) = \operatorname*{Res}_{s=0}\left[e^{st}F(s)\right] + \sum_{n=1}^{\infty}\operatorname*{Res}_{s=-n^2\pi^2}\left[e^{st}F(s)\right],$$

we arrive at the expression

$$f(t) = \frac{1}{6}x(x^2 - 1) + xt + \frac{2}{\pi^3}\sum_{n=1}^{\infty}\frac{(-1)^{n+1}}{n^3}e^{-n^2\pi^2 t}\sin n\pi x.$$

10. The function

$$F(s) = \frac{1}{s^2} - \frac{1}{s\sinh s}$$

has isolated singularities at the points

$$s_0 = 0 \quad \text{and} \quad s_n = n\pi i, \ \overline{s}_n = -n\pi i \ (n = 1, 2, \ldots).$$

Now

$$s\sinh s = s\left(s+\frac{1}{6}s^3+\cdots\right) = s^2+\frac{1}{6}s^4+\cdots \qquad (0<|s|<\infty),$$

and division of this series into 1 reveals that

$$F(s) = \frac{1}{s^2}-\left(\frac{1}{s^2}+\frac{1}{6}+\cdots\right) = -\frac{1}{6}+\cdots \qquad (0<|s|<\pi).$$

This shows that $F(s)$ has a removable singularity at s_0. Evidently, then, $e^{st}F(s)$ must also have a removable singularity there; and so

$$\operatorname*{Res}_{s=s_0}\left[e^{st}F(s)\right] = 0.$$

To find the residue of $F(s)$ at $s_n = n\pi i\ (n=1,2,\ldots)$, we write

$$F(s) = \frac{p(s)}{q(s)} \quad \text{where} \quad p(s) = \sinh s - s \ \text{ and } \ q(s) = s^2\sinh s$$

and observe that

$$p(n\pi i) = -n\pi i \neq 0, \quad q(n\pi i) = 0, \quad \text{and} \quad q'(n\pi i) = n^2\pi^2(-1)^{n+1} \neq 0.$$

Consequently, $F(s)$ has a simple pole at s_n, and

$$\operatorname*{Res}_{s=s_n} F(s) = \frac{p(n\pi i)}{q'(n\pi i)} = \frac{-n\pi i}{n^2\pi^2(-1)^{n+1}} = \frac{(-1)^n}{n\pi}i \ (n=1,2,\ldots).$$

Since $\overline{F(s)} = F(\overline{s})$, the points $\overline{s}_n$ are also simple poles of $F(s)$; and we may write

$$\operatorname*{Res}_{s=s_n}\left[e^{st}F(s)\right] + \operatorname*{Res}_{s=\overline{s}_n}\left[e^{st}F(s)\right] = 2\operatorname{Re}\left[\frac{(-1)^n}{n\pi}ie^{in\pi t}\right] = 2\operatorname{Re}\left[\frac{(-1)^n}{n\pi}(i\cos n\pi t - \sin n\pi t)\right]$$

$$= 2\frac{(-1)^{n+1}}{n\pi}\sin n\pi t.$$

Hence the desired result is

$$f(t) = \operatorname*{Res}_{s=s_0}\left[e^{st}F(s)\right] + \sum_{n=1}^{\infty}\left\{\operatorname*{Res}_{s=s_n}\left[e^{st}F(s)\right] + \operatorname*{Res}_{s=\overline{s}_n}\left[e^{st}F(s)\right]\right\},$$

or

$$f(t) = \frac{2}{\pi}\sum_{n=1}^{\infty}\frac{(-1)^{n+1}}{n}\sin n\pi t.$$

11. We consider here the function

$$F(s) = \frac{\sinh(xs)}{s(s^2+\omega^2)\cosh s} \qquad (0 < x < 1),$$

where $\omega > 0$ and $\omega \neq \omega_n = \dfrac{(2n-1)\pi}{2}$ $(n = 1,2,\ldots)$. The singularities of $F(s)$ are at

$$s = 0, \quad s = \pm\omega i, \quad \text{and} \quad s = \pm\omega_n i \ (n = 1,2,\ldots).$$

Because the first term in the Maclaurin series for $\sinh(xs)$ is xs, it is easy to see that $s = 0$ is a removable singularity of $e^{st}F(s)$ and that

$$\operatorname*{Res}_{s=s_0}\left[e^{st}F(s)\right] = 0.$$

To find the residue of $F(s)$ at $s = \omega i$, we write

$$F(s) = \frac{\phi(s)}{s-\omega i} \quad \text{where} \quad \phi(s) = \frac{\sinh(xs)}{s(s+\omega i)\cosh s},$$

from which it follows that $s = \omega i$ is simple pole and

$$\operatorname*{Res}_{s=\omega i} F(s) = \phi(\omega i) = \frac{\sinh(x\omega i)}{\omega i 2\omega i \cosh(\omega i)} = \frac{i\sin\omega x}{-2\omega^2\cos\omega}.$$

Since $\overline{F(s)} = F(\bar{s})$, then,

$$\operatorname*{Res}_{s=\omega i}\left[e^{st}F(s)\right] + \operatorname*{Res}_{s=-\omega i}\left[e^{st}F(s)\right] = 2\operatorname{Re}\left[\frac{i\sin\omega x}{-2\omega^2\cos\omega} ie^{i\omega t}\right] = 2\frac{\sin\omega x}{2\omega^2\cos\omega}\sin\omega t = \frac{\sin\omega x\sin\omega t}{\omega^2\cos\omega}.$$

As for the residues at $s = \omega_n i$ $(n = 1,2,\ldots)$, we put $F(s)$ in the form

$$F(s) = \frac{p(s)}{q(s)} \quad \text{where} \quad p(s) = \sinh(xs) \ \text{ and } \ q(s) = (s^3 + \omega^2 s)\cosh s.$$

Now $p(\omega_n i) = \sinh(x\omega_n i) = i\sin\omega_n x \neq 0$ and $q(\omega_n i) = 0$. Also, since

$$q'(s) = (s^3 + \omega^2 s)\sinh s + (3s^2 + \omega^2)\cosh s,$$

we find that

$$q'(\omega_n i) = (-\omega_n^3 i + \omega^2\omega_n i)\sinh(\omega_n i) = -\omega_n(\omega^2 - \omega_n^2)\sin\omega_n \neq 0.$$

Hence we have a simple pole at $s = \omega_n i$, with residue

$$\operatorname*{Res}_{s=\omega_n i} F(s) = \frac{p(\omega_n i)}{q'(\omega_n i)} = \frac{i\sin\omega_n x}{-\omega_n(\omega^2 - \omega_n^2)\sin\omega_n}.$$

Consequently,

$$\operatorname*{Res}_{s=\omega_n i}\left[e^{st}F(s)\right]+\operatorname*{Res}_{s=-\omega_n i}\left[e^{st}F(s)\right]=2\,\mathrm{Re}\left[\frac{i\sin\omega_n x}{-\omega_n(\omega^2-\omega_n^2)\sin\omega_n}e^{i\omega_n t}\right]=2\frac{\sin\omega_n x\sin\omega_n t}{\omega_n(\omega^2-\omega_n^2)\sin\omega_n}.$$

But $\sin\omega_n=\sin\left(n\pi-\frac{\pi}{2}\right)=(-1)^{n+1}$, and this means that

$$\operatorname*{Res}_{s=\omega_n i}\left[e^{st}F(s)\right]+\operatorname*{Res}_{s=-\omega_n i}\left[e^{st}F(s)\right]=2\frac{(-1)^{n+1}}{\omega_n}\cdot\frac{\sin\omega_n x\sin\omega_n t}{\omega^2-\omega_n^2}.$$

Finally,

$$f(t)=\operatorname*{Res}_{s=0}\left[e^{st}F(s)\right]+\left\{\operatorname*{Res}_{s=\omega i}\left[e^{st}F(s)\right]+\operatorname*{Res}_{s=-\omega i}\left[e^{st}F(s)\right]\right\}+\sum_{n=1}^{\infty}\left\{\operatorname*{Res}_{s=\omega_n i}\left[e^{st}F(s)\right]+\operatorname*{Res}_{s=-\omega_n i}\left[e^{st}F(s)\right]\right\}.$$

That is,

$$f(t)=\frac{\sin\omega x\sin\omega t}{\omega^2\cos\omega}+2\sum_{n=1}^{\infty}\frac{(-1)^{n+1}}{\omega_n}\cdot\frac{\sin\omega_n x\sin\omega_n t}{\omega^2-\omega_n^2}.$$